AUGER ELECTRON SPECTROSCOPY REFERENCE MANUAL

W0037696

AUGER ELECTRON SPECTROSCOPY REFERENCE MANUAL

A Book of Standard Spectra for Identification and Interpretation of Auger Electron Spectroscopy Data

G. E. McGuire

Texas Instruments, Inc.
Dallas, Texas

SPRINGER SCIENCE+BUSINESS MEDIA, LLC

Library of Congress Cataloging in Publication Data

McGuire, G   E
Auger electron spectroscopy reference manual.

Includes index.
1. Auger effect—Handbooks, manuals, etc. 2. Photoelectron spectro-
scopy—Handbooks, manuals, etc. I. Title.
QC793.5.E627M3   539.7'2112   79-24223
ISBN 978-1-4757-1704-4   ISBN 978-1-4757-1702-0 (eBook)
DOI 10.1007/978-1-4757-1702-0

© 1979 Springer Science+Business Media New York
Originally published by Plenum Press, New York in 1979
Softcover reprint of the hardcover 1st edition 1979

All rights reserved

No part of this book may be reproduced, stored in a retrieval system, or transmitted,
in any form or by any means electronic, mechanical, photocopying, microfilming,
recording, or otherwise, without written permission from the Publisher

INTRODUCTION

Auger electron spectroscopy (AES) is based on the Auger process, which involves the core-level ionization of an atom with subsequent deexcitation occurring by an outer-level electron decaying to fill the core hole. The excess energy is transferred to and causes the ejection of another electron, which is by definition an Auger electron. The Auger electron transition, denoted by the electron levels involved, is independent of the excitation source and leaves the atom with a constant kinetic energy. The kinetic energy is given by the differences in binding energies for the three levels (for example, $E_K - E_{L_1} - E_{L_{2,3}}$) minus a correction term for the work function and electron wave function relaxation. When the Auger transition occurs within a few angstroms of the surface, the Auger electrons may be ejected from the surface without loss of energy and give rise to peaks in the secondary electron energy distribution. Each element has a unique set of Auger transitions which may be used to identify the composition of solid surfaces.

By convention, Auger electron spectroscopy refers to electron excitation of the atom although Auger processes can be caused by incident photons, electrons, or ions. The basic Auger spectrometer consists of an ultrahigh-vacuum system to provide a contamination-free surface, an electron gun for specimen excitation, an electron analyzer to distinguish Auger electrons in the total secondary electron energy distribution, and an ion gun to provide depth profiling capability.

The high surface sensitivity of Auger spectroscopy which dictates the need for an ultrahigh-vacuum system is due to the limited mean free path of electrons in the 0–3000 eV kinetic energy range. The Auger peaks decay exponentially with overlayer coverage, which is consistent with an exponential dependence of escape probability on the depth of the parent atom. A compilation of data from a variety of sources has been used to generate an escape depth curve which falls in the range of 5–30 Å in the energy range from 0 to 3000 eV. The observed escape depth does not show a strong dependence on the matrix.

Electron beams provide a versatile excitation source. Electron beams can be varied in energy and focused to a small beam diameter with conventional deflection and rastering capability to obtain two-dimensional compositional analysis of surfaces. Since the Auger transition probability and Auger electron escape depth are independent of the excitation source, the dependence of the Auger peak amplitude on the incoming electron beam is governed by the ionization cross section of the initial core level. Ionization occurs primarily by the incident electrons, but backscattered primary electrons can also contribute to the Auger yield when the incident beam energy is substantially greater than the binding energy of the core level involved. The Auger yield rises above zero above the ionization threshold and increases to a maximum when the primary beam energy is three to five times greater than the core energy level.

In the development of Auger spectroscopy for surface analysis several types of electron energy analyzers have been employed. Because of its superior signal-to-noise ratio, the cylindrical mirror analyzer with a coaxial electron gun is used almost exclusively with modern Auger spectrometers. The optical axis of the electron gun is coincident with the cylindrical mirror axis. The electron beam is focused to a fine point on the surface of the specimen, which is positioned at the source point of the analyzer. Electrons ejected from the point of excitation pass through a grid-covered aperture on the inner cylinder. A negative potential applied to the outer cylinder deflects electrons with proper energy originating from the sample within the focal point of the analyzer through a second aperture on the inner cylinder, and finally through a small exit aperture to the detector. By varying the potential applied to the outer cylinder, a range of electron energies may be examined. Most commercial Auger spectrometers are capable of approximately 0.5% energy resolution, $\Delta E/E$, with 10% transmission.

The secondary electron energy distribution, $N(E)$, is generated by plotting the output of the electron multiplier versus the negative voltage applied to the outer cylinder. Since the Auger electron intensity is small and is superimposed upon the high background caused by inelastically scattered electrons, Auger electron spectra are normally taken in the derivative mode. The derivative $dN(E)/dE$ is obtained by superimposing a small sinusoidal potential modulation on the analyzer pass energy and synchronously detecting the current passed through the analyzer. It is a common practice in electron-excited AES to use the peak-to-

peak signal strength in the derivative spectrum as a relative quantitative measure of elemental surface concentration. In addition, the energy positions of the negative-going peaks in the derivative spectrum are used to identify the Auger transition energy values. One judgment that must be made by the user of AES is the necessary tradeoff between sensitivity and resolution in setting the amplitude of the potential modulation employed in electronic differentiation. By electronically varying the modulation voltage, one has versatile control over the signal-to-noise ratio. Information of the lowest order concerning an Auger feature is available if one is interested only in detecting a signal. This is limited by shot noise associated with the background current upon which the Auger peaks are superimposed to a range of 100 to 1000 ppm. The signal strength is optimum when the modulation voltage matches the natural line width of the Auger peaks.

The AES features actually contain more detail than is frequently utilized. The structures usually consist of a main peak followed by additional features on the low-energy side because of various couplings of the Auger transition to the valence band electrons. Higher energy resolution, or low modulation voltage, can yield fine structure in Auger peaks which depend on the chemical environment of the atoms being studied, but will result in a significant reduction in signal-to-noise performance for many Auger peaks. However, the use of large modulation voltages is beneficial when quantitative measurements are affected by the primary electron beam current, as the improved signal-to-noise ratio can be traded off for either faster energy analysis or a reduction in primary electron beam current. Since the first harmonic signal strength is nonlinear, it is necessary to know the signal strength of the characteristic Auger transitions for various elements as a function of modulation voltage.

The inherent surface sensitivity of AES may be utilized in combination with ion sputtering to obtain depth information in thin-film analysis. A sputter ion gun is operated simultaneously with the electron gun to ion-beam-mill a crater which is large compared with the diameter of the electron beam probe. Depth information is acquired by continually monitoring the elemental composition of the crater bottom during sputtering erosion. In typical profile measurements, the surface is sputtered away at a rate of several atomic layers per second under a static pressure of 3.8×10^{-3} pascal argon. Ion beam uniformity across the sampled area, the Auger electron escape depth, and sample homogeneity affect the depth resolution of AES. Loss of depth resolution from ion beam nonuniformity is negligible if the ion beam is large compared with the electron beam. The Auger escape depth contribution remains constant at 5–20 Å. The best depth resolution is achieved in amorphous films, where the sputter rate is not affected by grain orientation, precipitates, or impurities.

Auger electron transitions require three electron levels so that only elements with atomic numbers greater than three can be detected. The rate of core-level ionization, being one of the key factors in Auger transition intensity, can be adjusted by varying the primary electron beam energy so that the relative KLL, LMM, and MNN intensities are altered. The KLL Auger transitions are the most intense for low-atomic-number elements, but the LMM transitions increase in intensity with increasing atomic number, and subsequently the MNN transitions increase as well. By progressively using the KLL, LMM, and MNN series of Auger transitions, the elemental sensitivity variation across the periodic table can be held to a factor of less than fifty.

A highly useful method for determining atomic concentrations makes use of the atomic KLL, LMM, and MNN transition intensities. Assuming that the transition intensities can be measured for the pure elements under a set of controlled conditions, the atomic concentration of element X can be expressed as

$$C_X = (I_X/S_X) \Big/ \sum_a (I_a/S_a)$$

where S_X is the relative sensitivity and I_X the Auger transition intensity of element X. The relative sensitivity of element X and a chosen standard can be obtained by

$$S_X = \left(\frac{A+B}{A} \right) \frac{I_X}{I_s}$$

where A and B are the chemical formula indices of compound $X_A Y_B$, and I_s is the Auger transition intensity of the standard. The method obviously neglects variations in the Auger yield to backscattered electrons, electron escape depth, and surface roughness or topography. Generally one assumes that surface topography affects all peaks uniformly. The method then gives semiquantitative results without the use of standards. Other, more quantitative, techniques may be developed through the use of external standards.

Experimental

The Auger results reported here were obtained with a single-pass cylindrical mirror analyzer (CMA) manufactured by Physical Electronics, Inc. (PHI Model 10-155). Samples were mounted at an angle of 30° with respect to the coincident 5 keV, 5 μA electron beam. Prior to analysis the sample surface was cleaned by 2 keV argon or xenon ion bombardment for a minimum of 15 minutes at a pressure of 3.8×10^{-3} pascal. Ion sputtering at normal incidence was continued during data analysis to ensure a clean surface. The energy scale was calibrated by measuring the analyzer voltage required to transmit elastically scattered 2 keV primary electrons. All spectra were recorded using a silver standard taken under a constant set of conditions. The multiplier gain was the only variable used to maintain a constant peak-to-peak amplitude for the 351 eV Ag transition.

Typically, the data are presented in three ways (though the latter two are omitted for some of the elements):

1. The general survey scans from 0 to 2000 eV were taken with a 5 keV primary beam energy, 5 μA beam current, and 6 eV peak-to-peak modulation voltage.

2. Selected characteristic transitions are displayed at 50 eV/division taken at higher resolution with a 1 eV peak-to-peak modulation voltage. The beam voltage and beam current remained the same as for the survey scan. The inserted graphs show the peak-to-peak intensity as a function of primary beam voltage for some of the key characteristic Auger transitions. In this case the modulation voltage was held constant at 6 eV.

3. In the next series of curves characteristic Auger transitions are displayed as a function of modulation voltage while maintaining a 5 μA beam current and 5 keV primary beam energy. The lock-in amplifier gain and the electron energy scales are all indicated on the figures.

The data in this handbook do not cover the entire periodic table, although all the most frequently encountered elements are included. The data give the analyst in the laboratory an opportunity to examine the spectral features at both high and low resolution for structural variations. Also, they give the variations in signal intensity as a function of primary beam energy and modulation voltage. The data will not universally match data from other analyzers because of slight variations in design or mechanical tolerances and in the focusing properties of the electron gun. As a result, truly quantitative analyses can only be achieved by running calibration standards at the time of analysis. The data can be used as a laboratory guide to the key instrumental parameters used in Auger electron spectroscopy.

Acknowledgments

The author would like to thank Bob Martin for the many hours he spent in obtaining the spectra in this compilation and Billy Davis for his skillful and timely efforts in preparing the artwork. Without their support this book would never have been completed.

CONTENTS

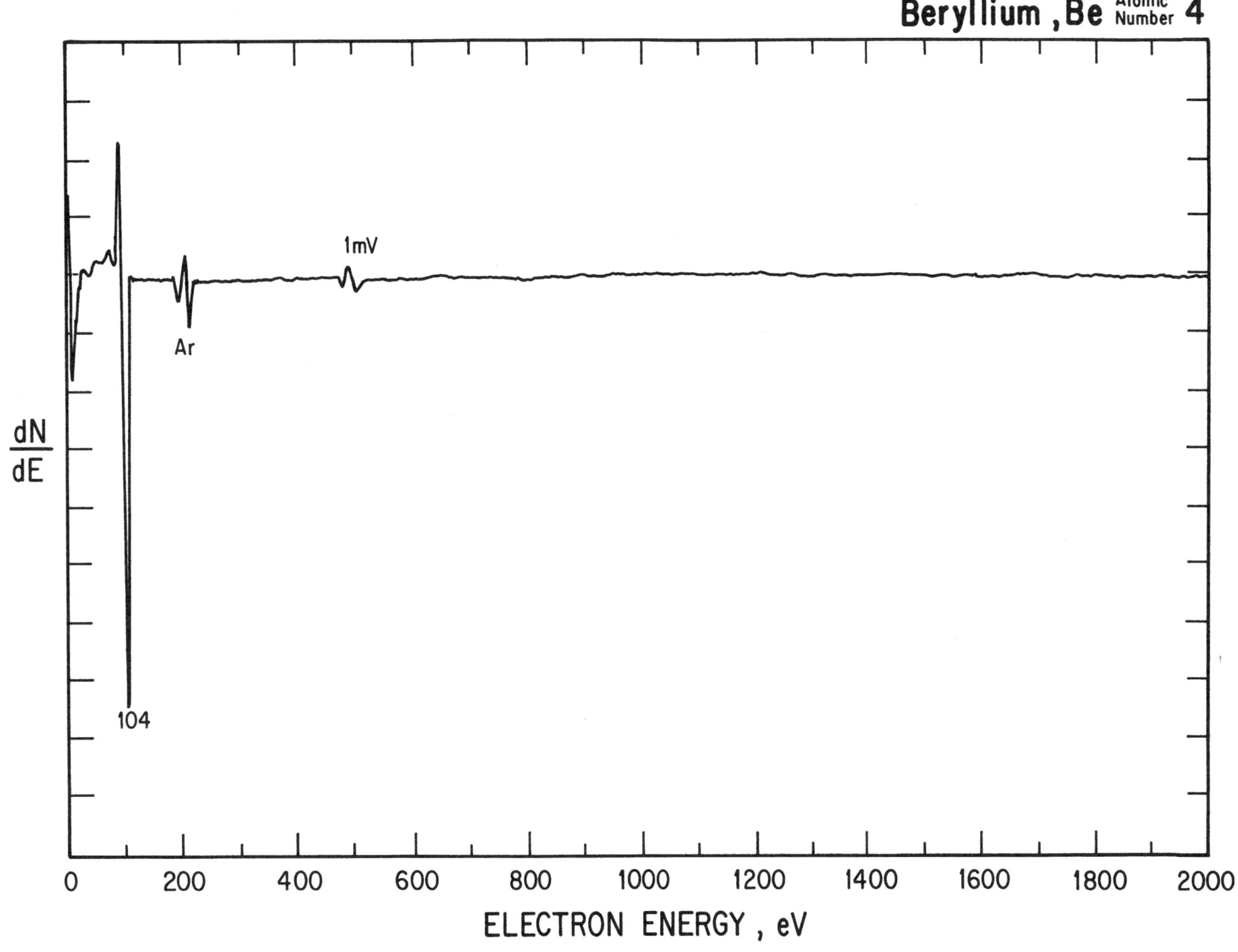

Beryllium, Be Atomic Number 4

$\dfrac{dN}{dE}$

ELECTRON ENERGY, eV

1mV

Ar

104

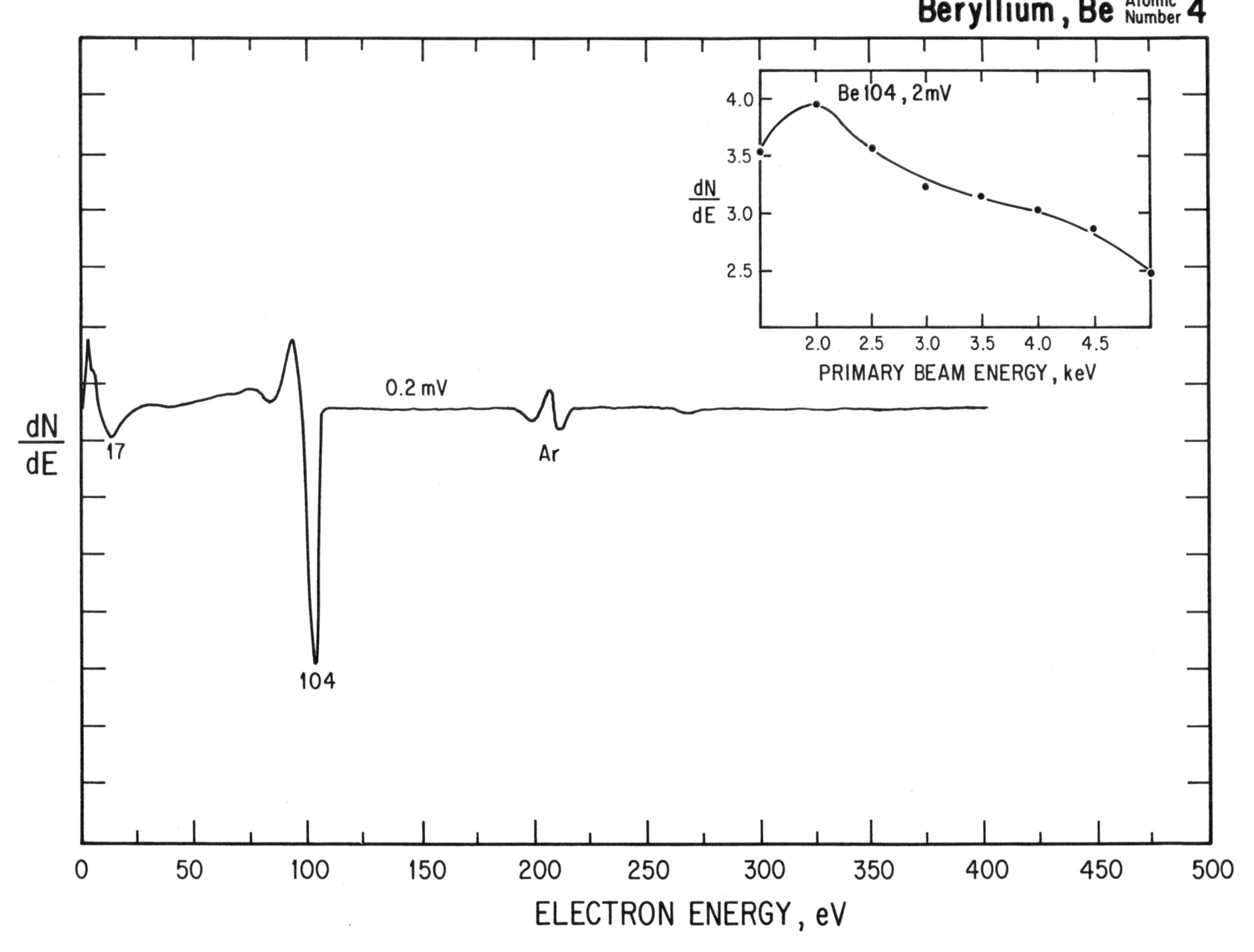

Beryllium, Be Atomic Number 4

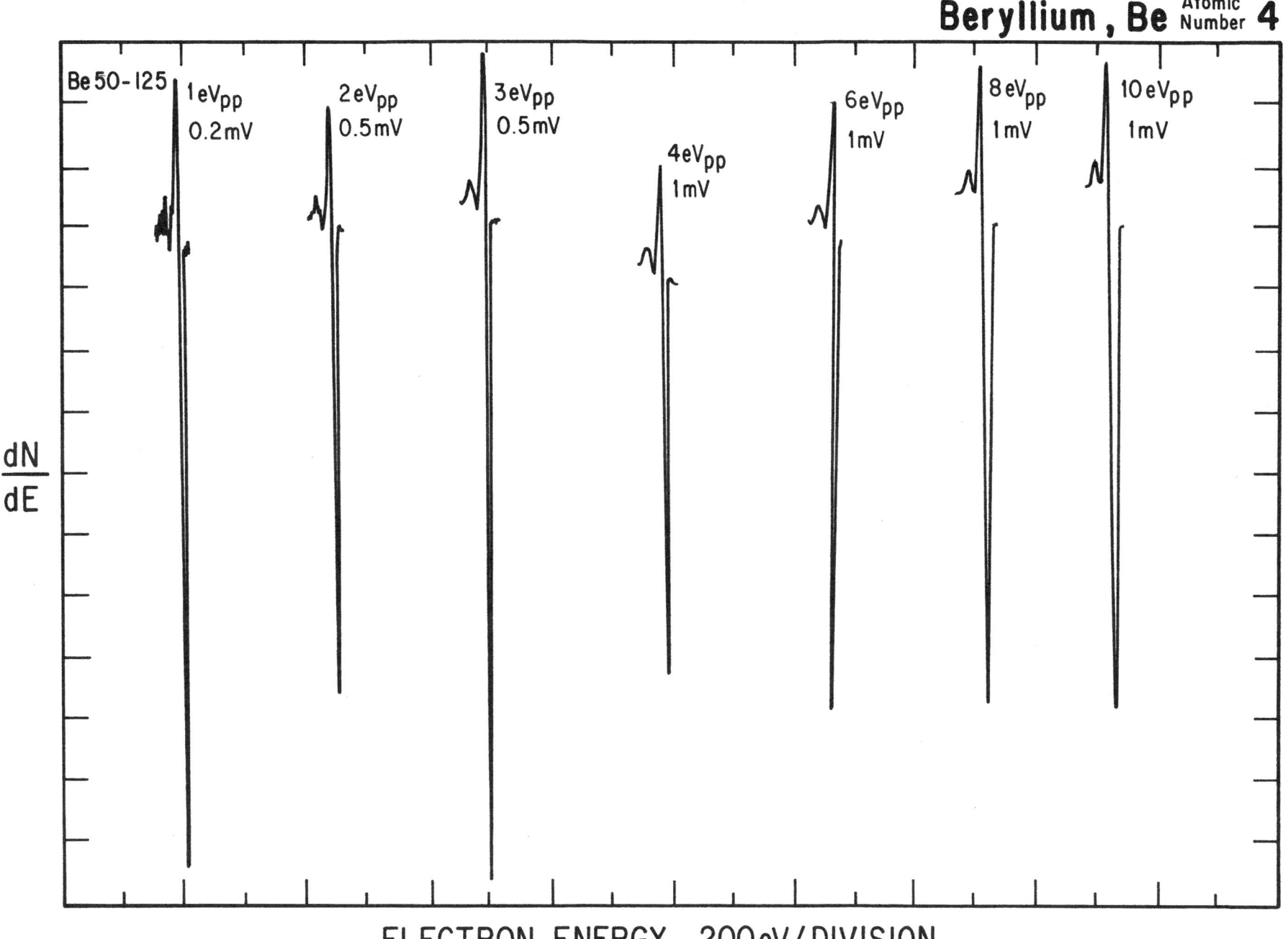

Beryllium, Be Atomic Number 4

Be 50-125   1 eV_pp   2 eV_pp   3 eV_pp   4 eV_pp   6 eV_pp   8 eV_pp   10 eV_pp
            0.2 mV   0.5 mV   0.5 mV   1 mV   1 mV   1 mV   1 mV

dN/dE

ELECTRON ENERGY, 200 eV/DIVISION

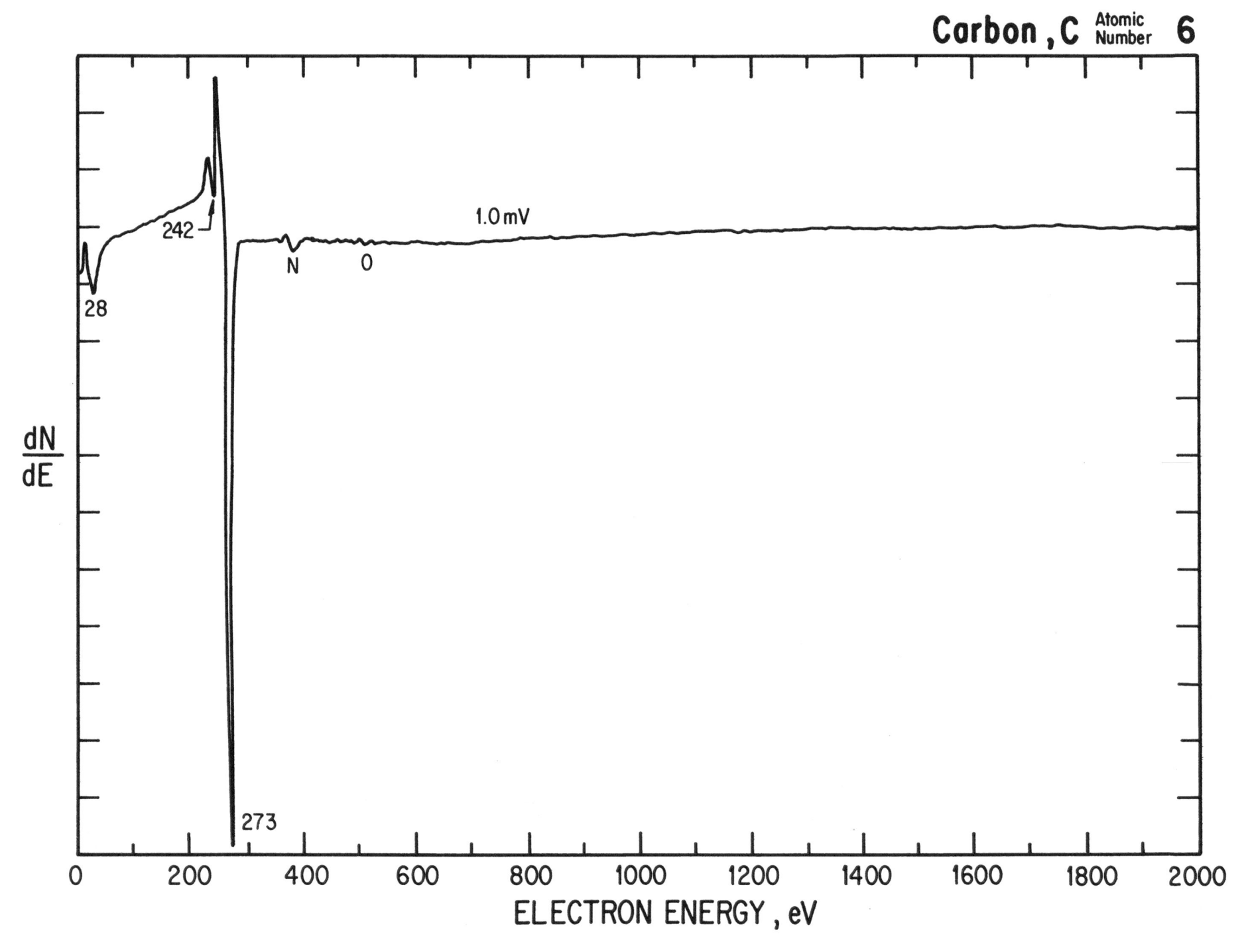

Carbon, C Atomic Number 6

1.0 mV

242

28

273

N

O

$\frac{dN}{dE}$

ELECTRON ENERGY, eV

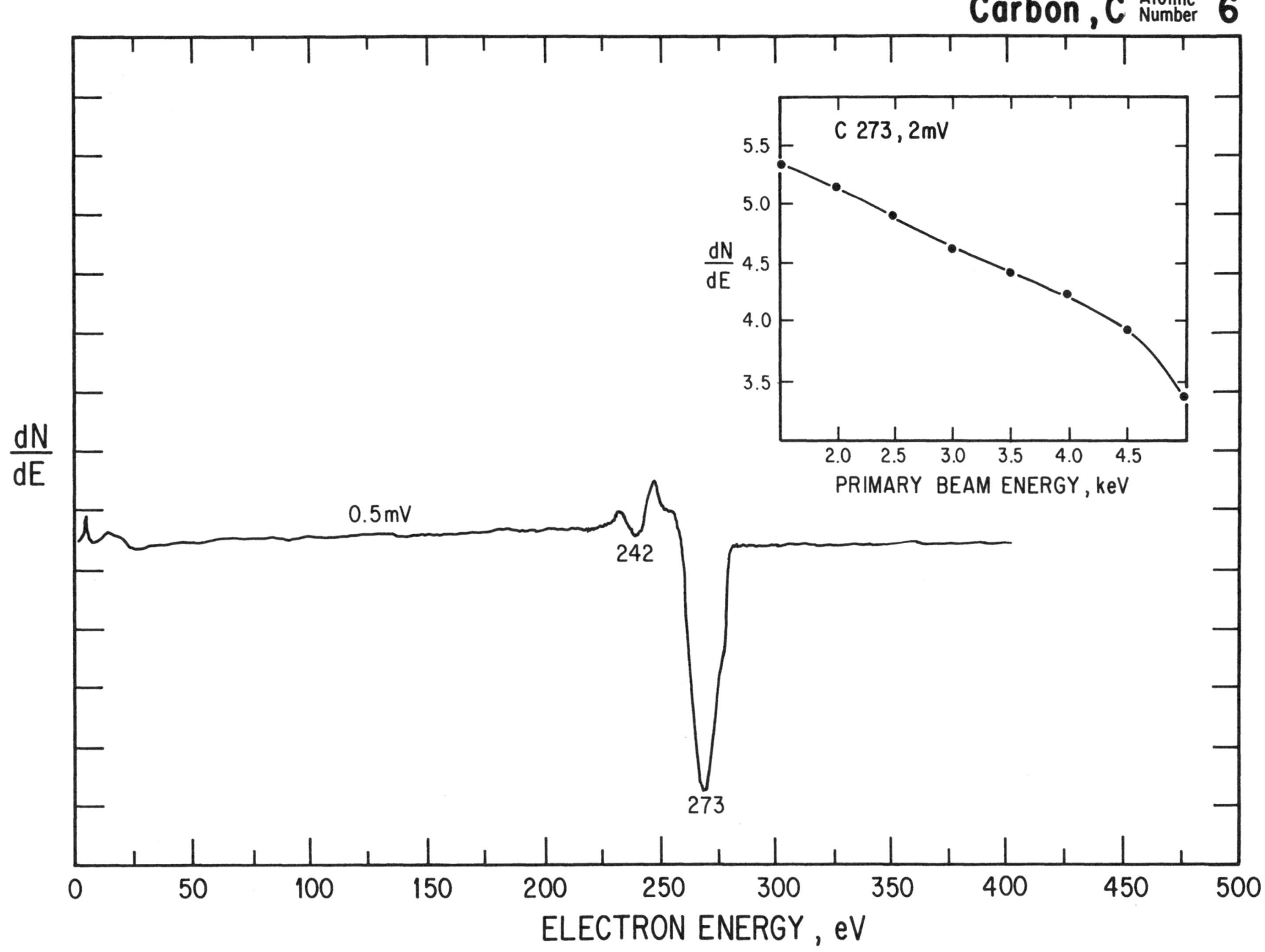

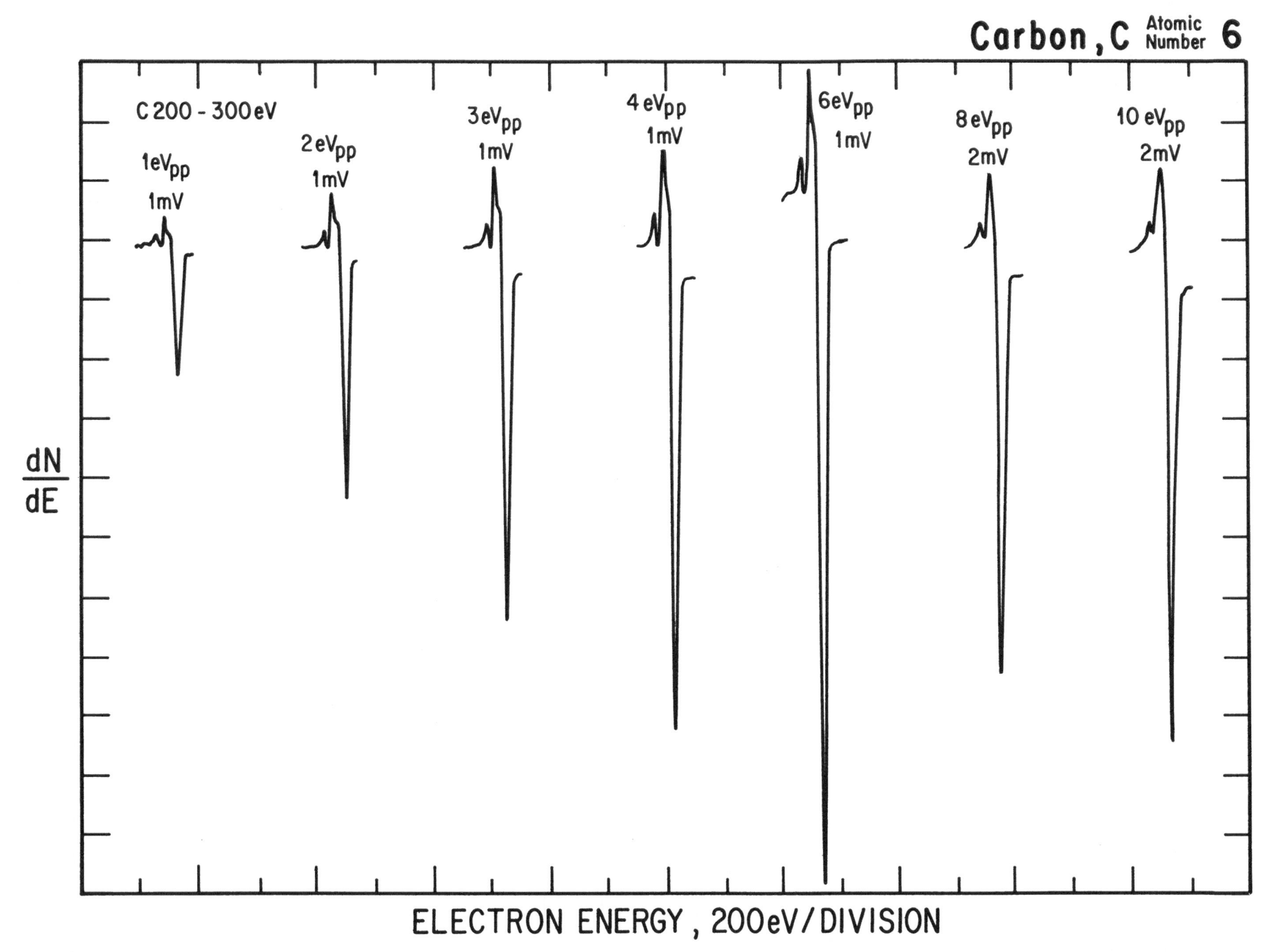

Carbon, C Atomic Number **6**

C 200 - 300 eV

1eV_pp 1mV

2eV_pp 1mV

3eV_pp 1mV

4eV_pp 1mV

6eV_pp 1mV

8eV_pp 2mV

10eV_pp 2mV

$\dfrac{dN}{dE}$

ELECTRON ENERGY, 200eV/DIVISION

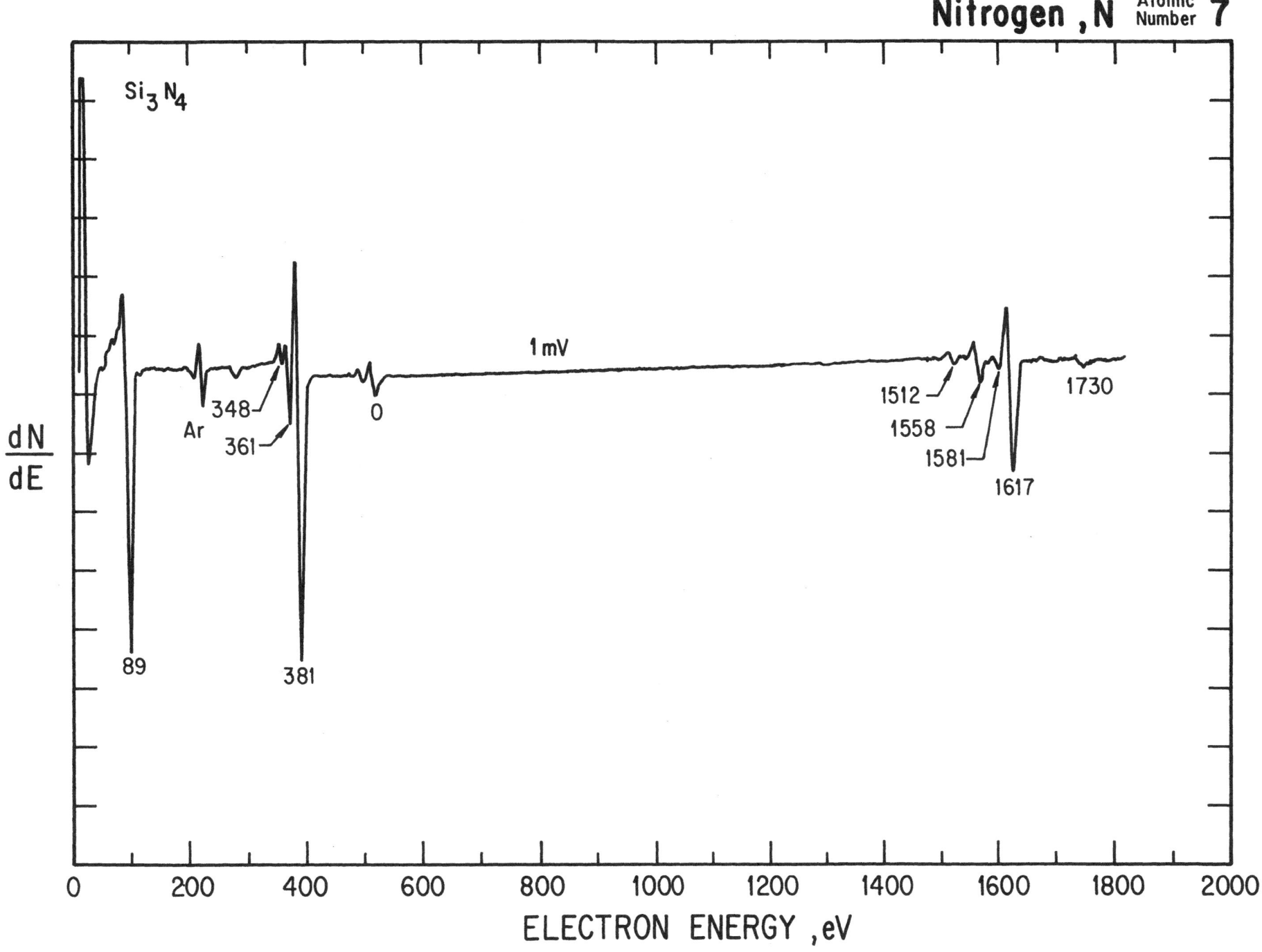

Si$_3$N$_4$

$\dfrac{dN}{dE}$

1 mV

348

Ar

361

0

1512

1558

1581

1617

1730

89

381

ELECTRON ENERGY , eV

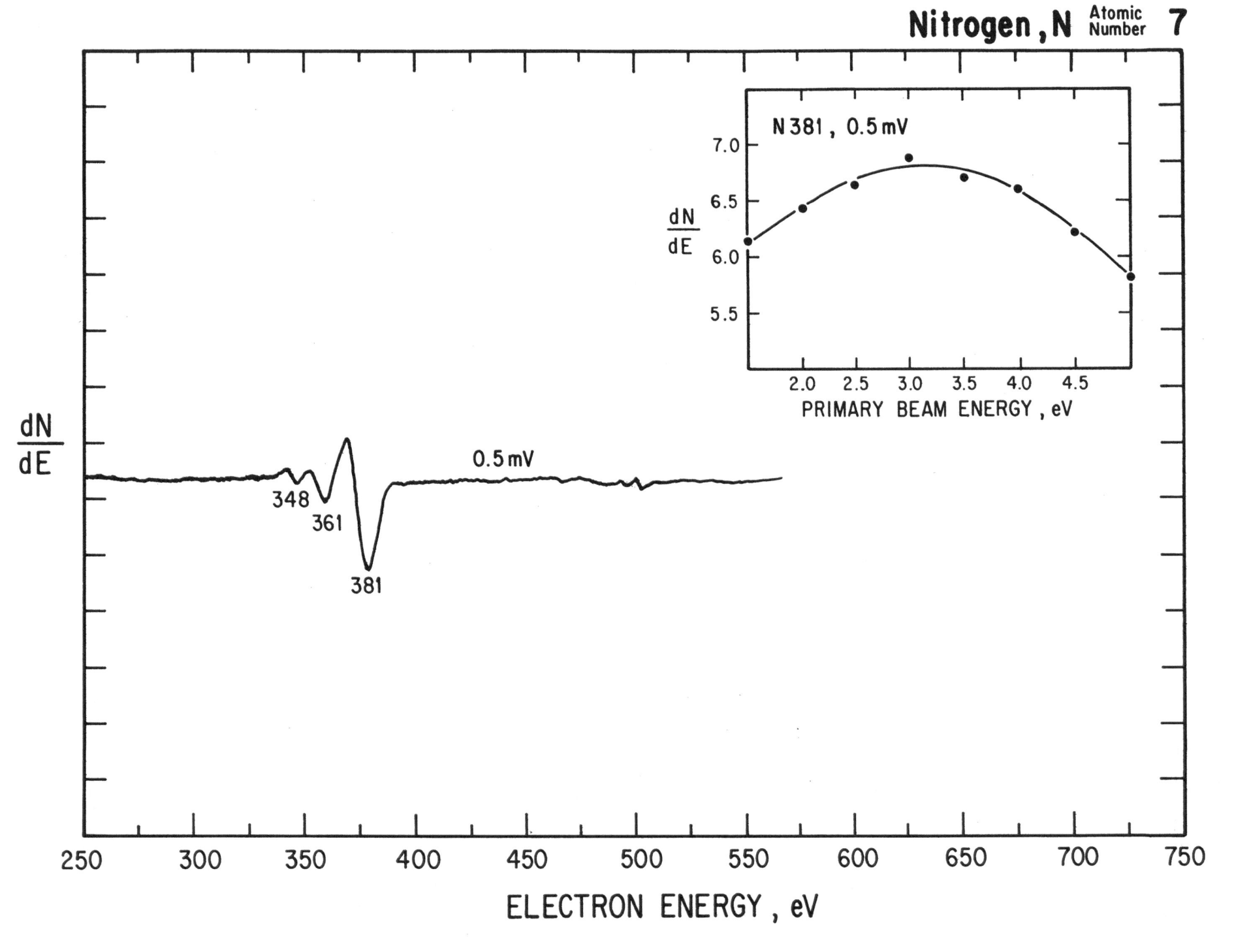

$\dfrac{dN}{dE}$

N 381, 0.5 mV

$\dfrac{dN}{dE}$

7.0

6.5

6.0

5.5

2.0   2.5   3.0   3.5   4.0   4.5

PRIMARY BEAM ENERGY, eV

0.5 mV

348

361

381

ELECTRON ENERGY, eV

250   300   350   400   450   500   550   600   650   700   750

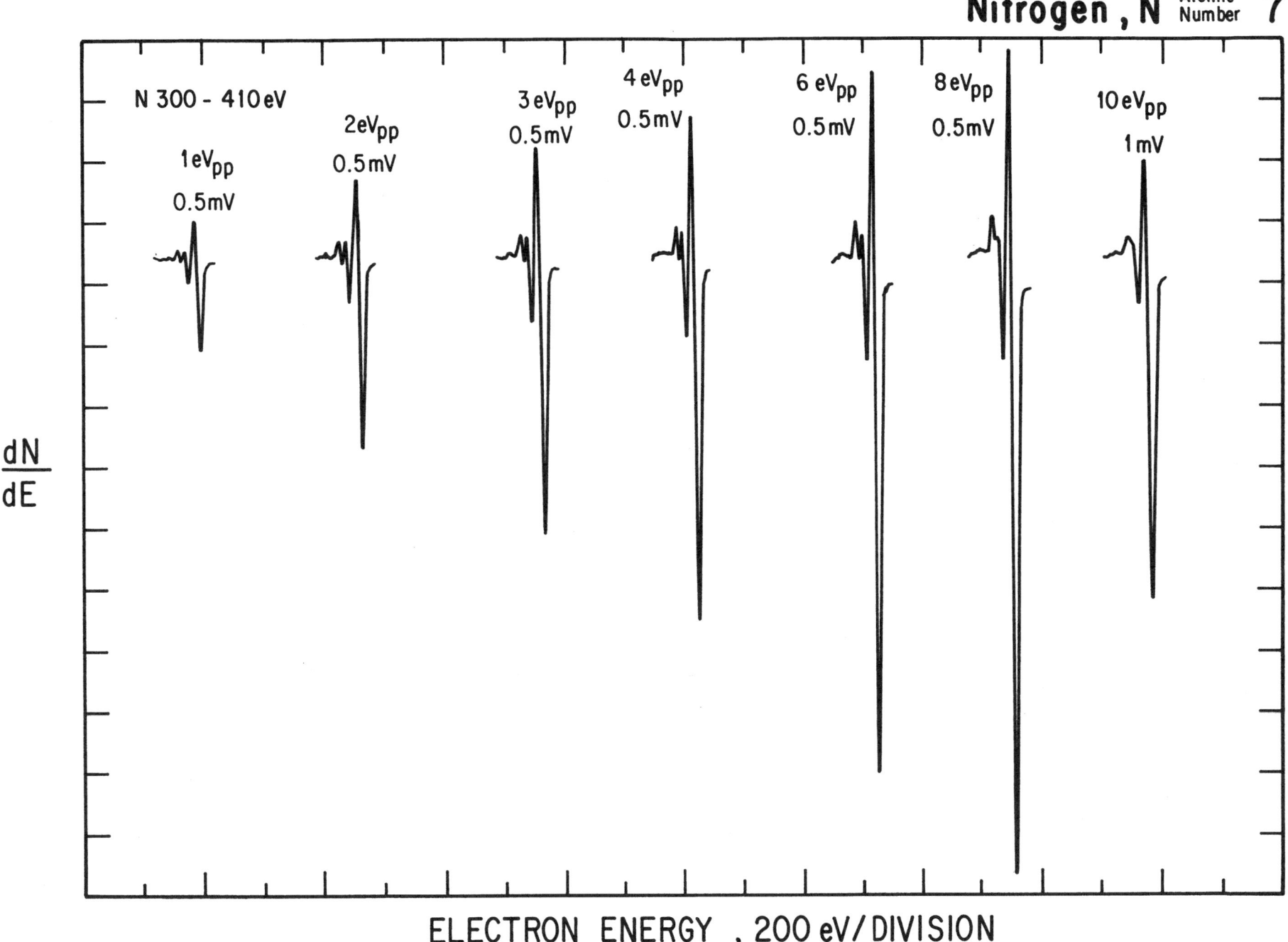

$\dfrac{dN}{dE}$

N 300 - 410 eV

1eV$_{pp}$
0.5mV

2eV$_{pp}$
0.5mV

3eV$_{pp}$
0.5mV

4 eV$_{pp}$
0.5mV

6 eV$_{pp}$
0.5mV

8eV$_{pp}$
0.5mV

10eV$_{pp}$
1 mV

ELECTRON ENERGY , 200 eV/DIVISION

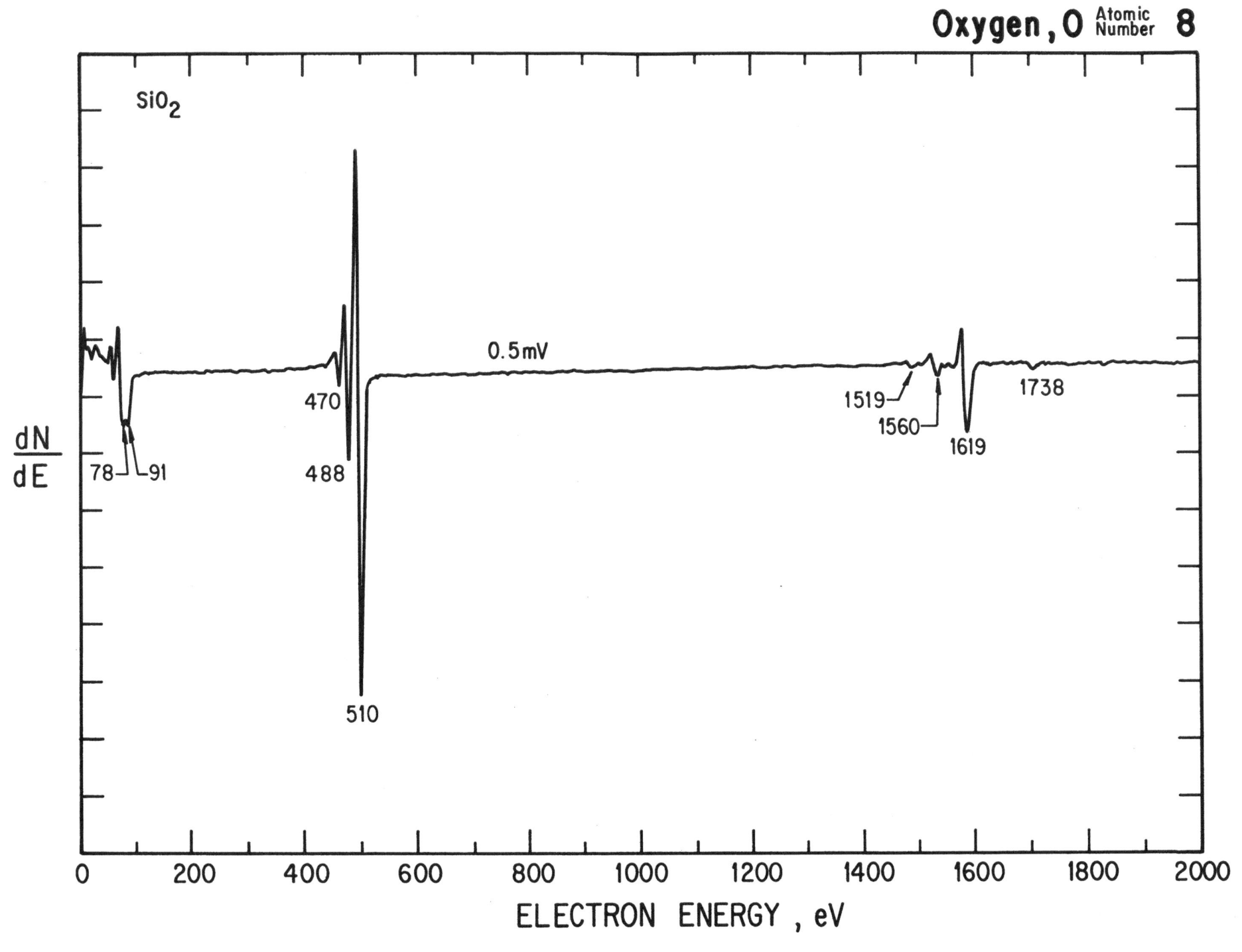

$\dfrac{dN}{dE}$

SiO$_2$

0.5 mV

78 — 91

470

488

510

1519

1560

1619

1738

ELECTRON ENERGY , eV

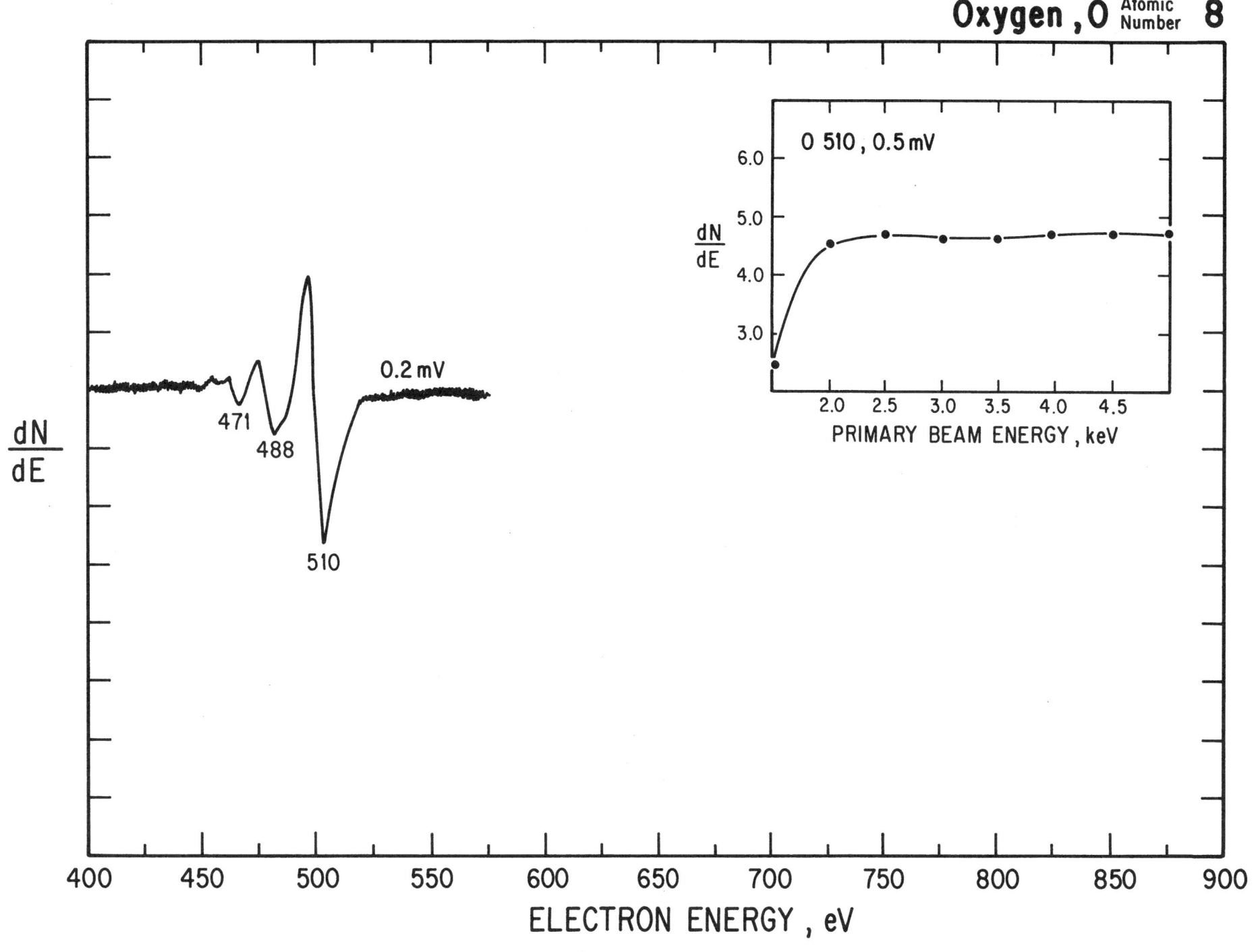

$\dfrac{dN}{dE}$

471
488
510

0.2 mV

O 510, 0.5 mV

$\dfrac{dN}{dE}$

6.0

5.0

4.0

3.0

2.0   2.5   3.0   3.5   4.0   4.5

PRIMARY BEAM ENERGY, keV

ELECTRON ENERGY, eV

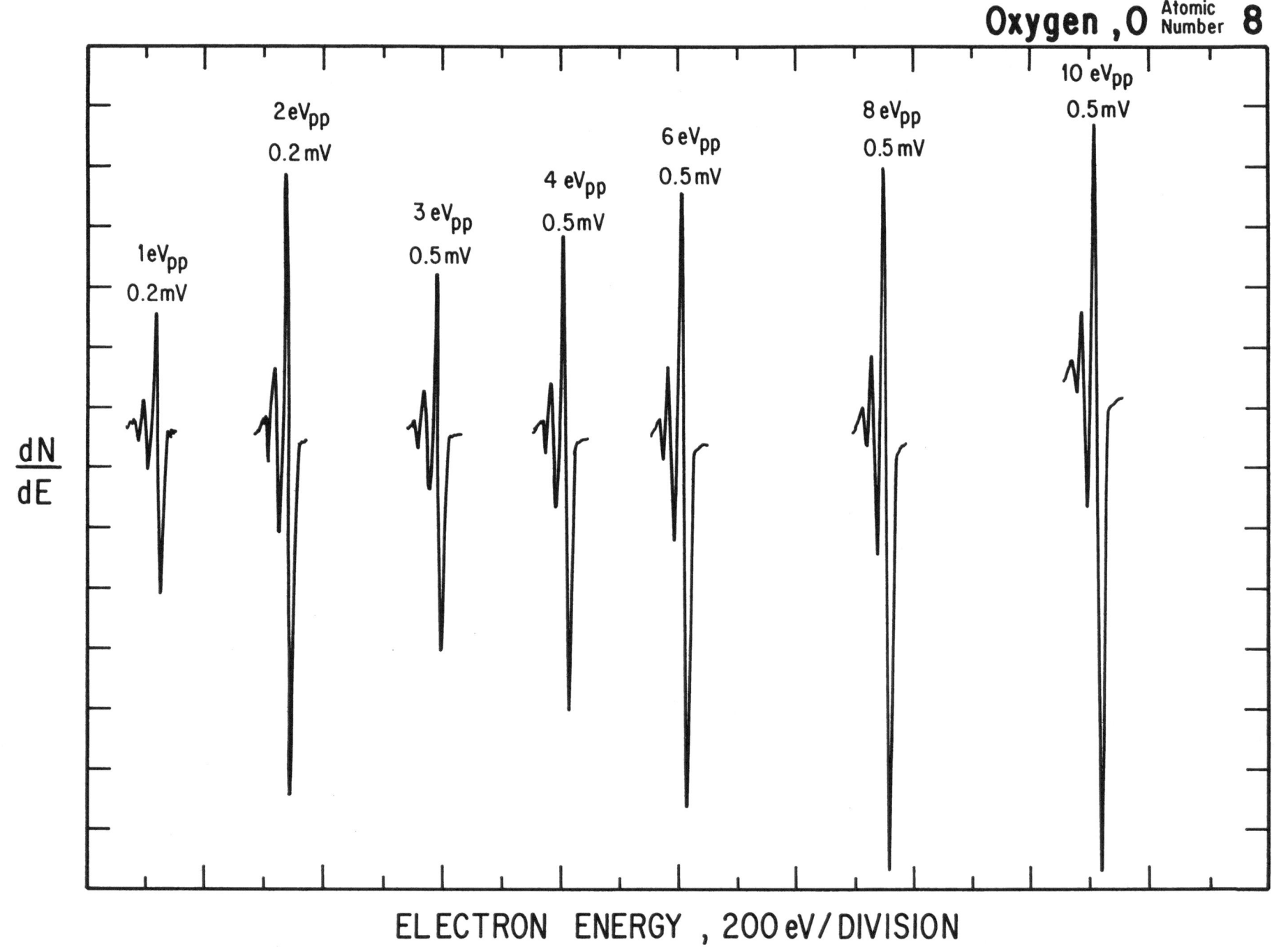

$\frac{dN}{dE}$

1eV_pp 0.2mV

2eV_pp 0.2 mV

3 eV_pp 0.5 mV

4 eV_pp 0.5mV

6 eV_pp 0.5 mV

8 eV_pp 0.5 mV

10 eV_pp 0.5mV

ELECTRON ENERGY , 200 eV/DIVISION

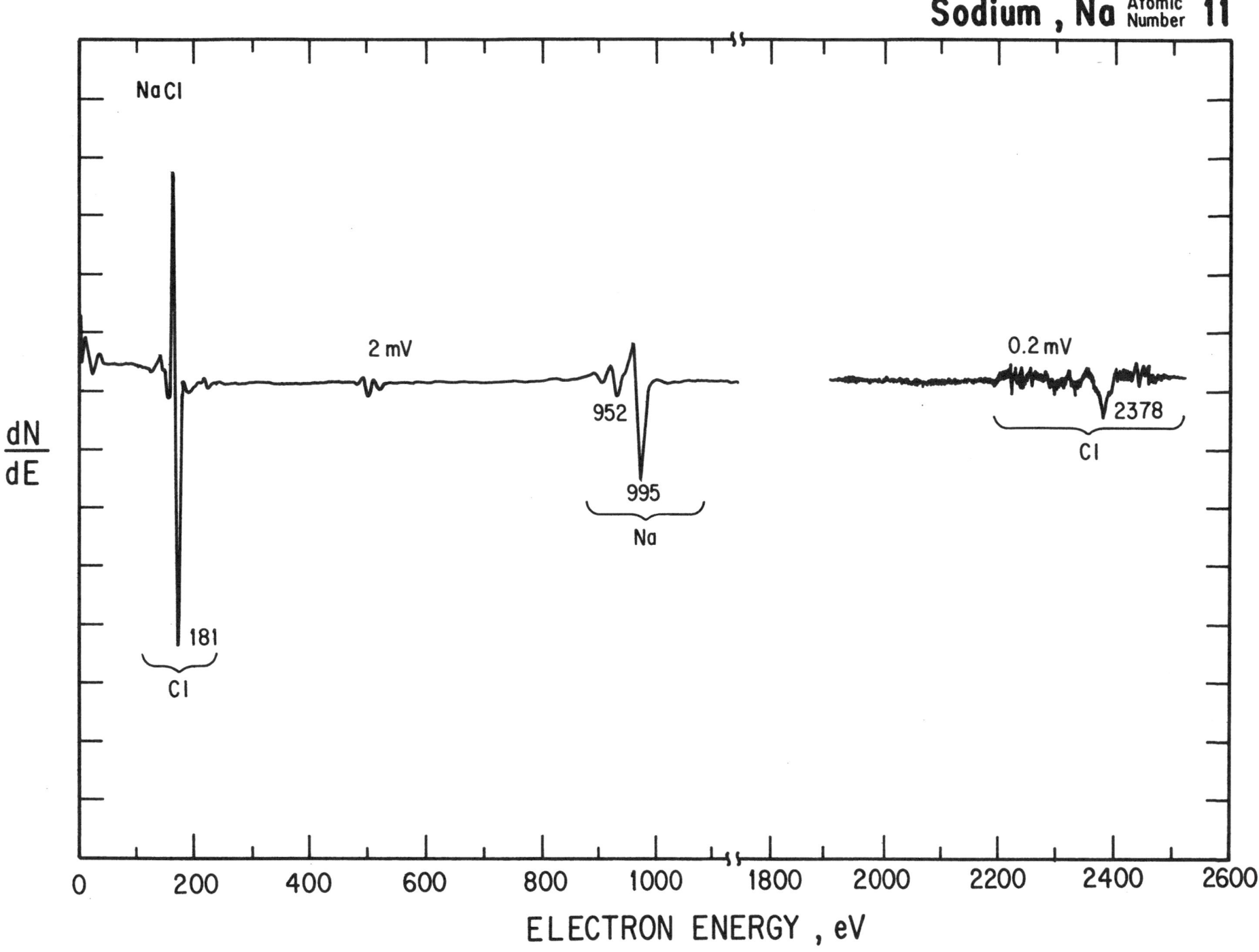

NaCl

2 mV

0.2 mV

$\dfrac{dN}{dE}$

952

995

Na

181

Cl

2378

Cl

ELECTRON ENERGY , eV

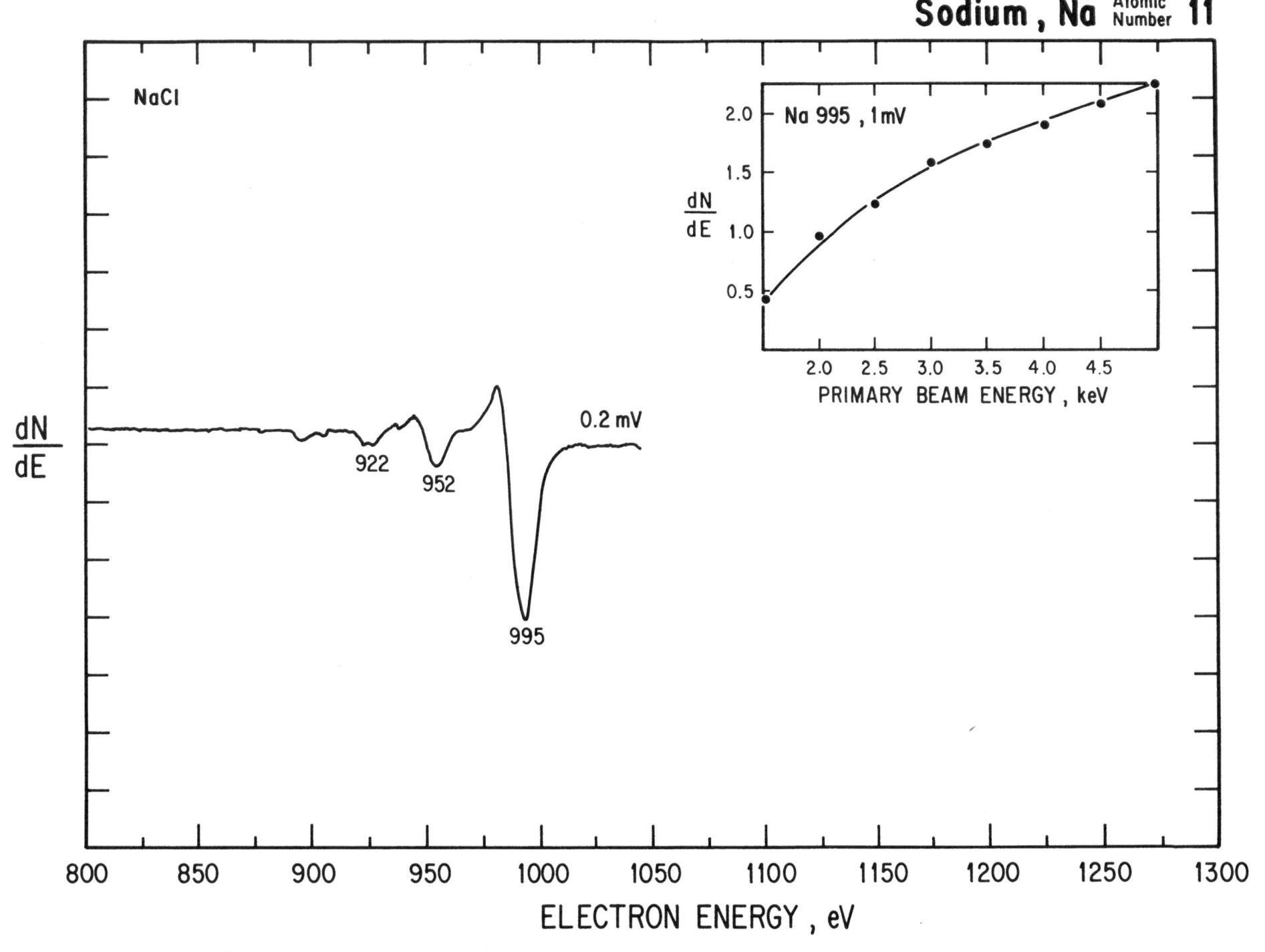

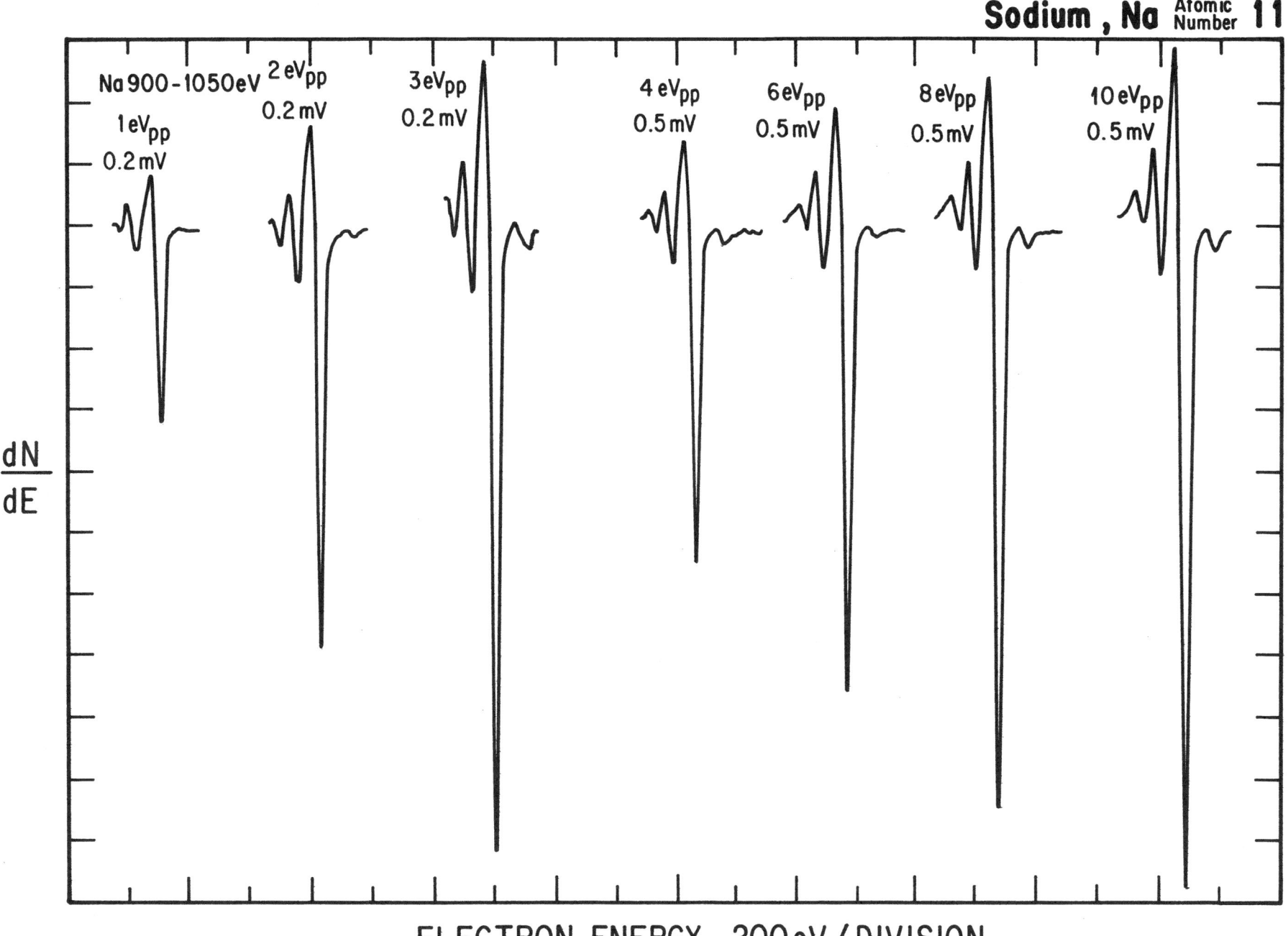

Sodium , Na Atomic Number **11**

Na 900-1050eV

1 eV$_{pp}$
0.2 mV

2 eV$_{pp}$
0.2 mV

3 eV$_{pp}$
0.2 mV

4 eV$_{pp}$
0.5 mV

6 eV$_{pp}$
0.5 mV

8 eV$_{pp}$
0.5 mV

10 eV$_{pp}$
0.5 mV

$\dfrac{dN}{dE}$

ELECTRON ENERGY , 200 eV / DIVISION

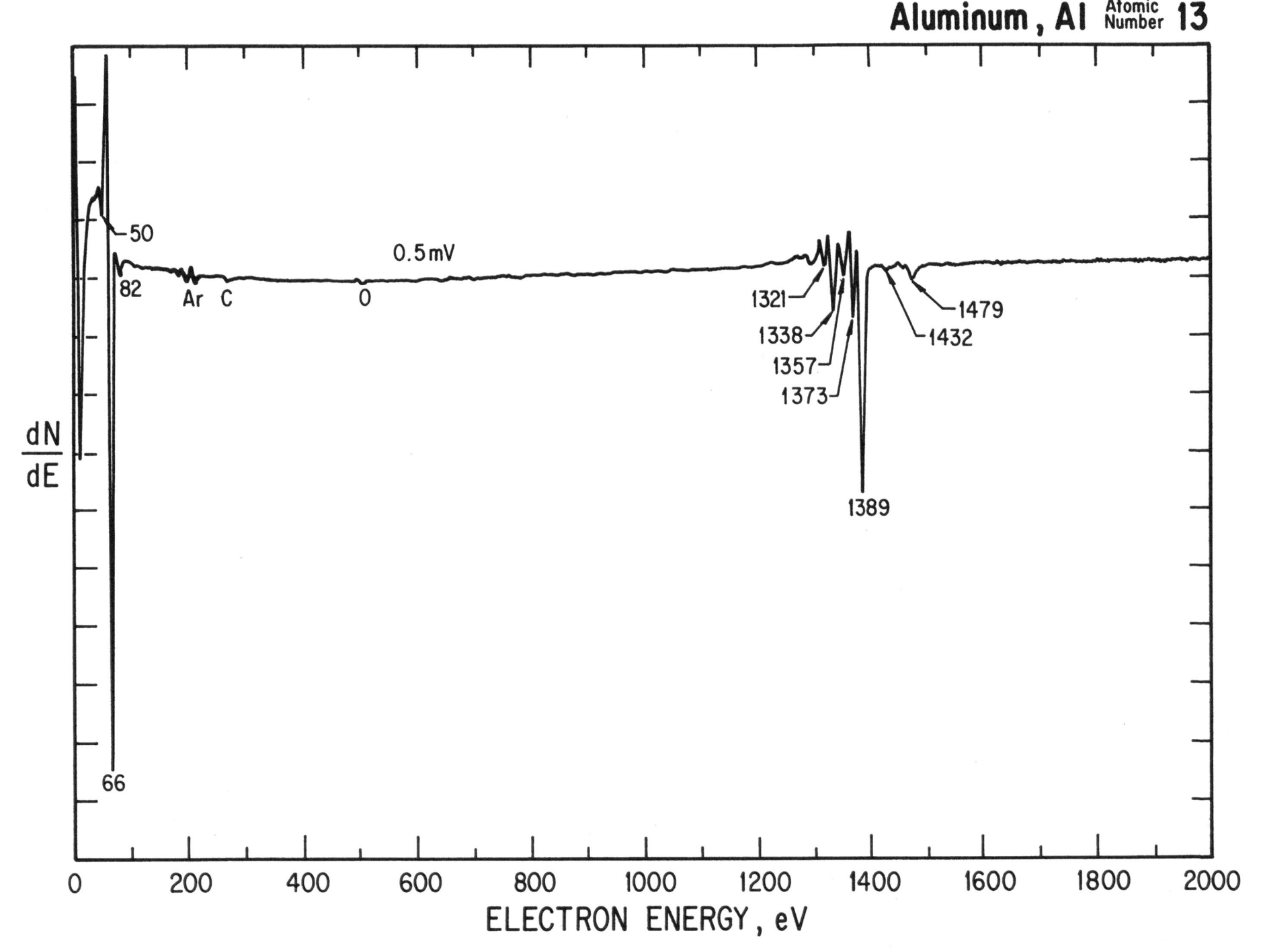

Aluminum, Al Atomic Number 13

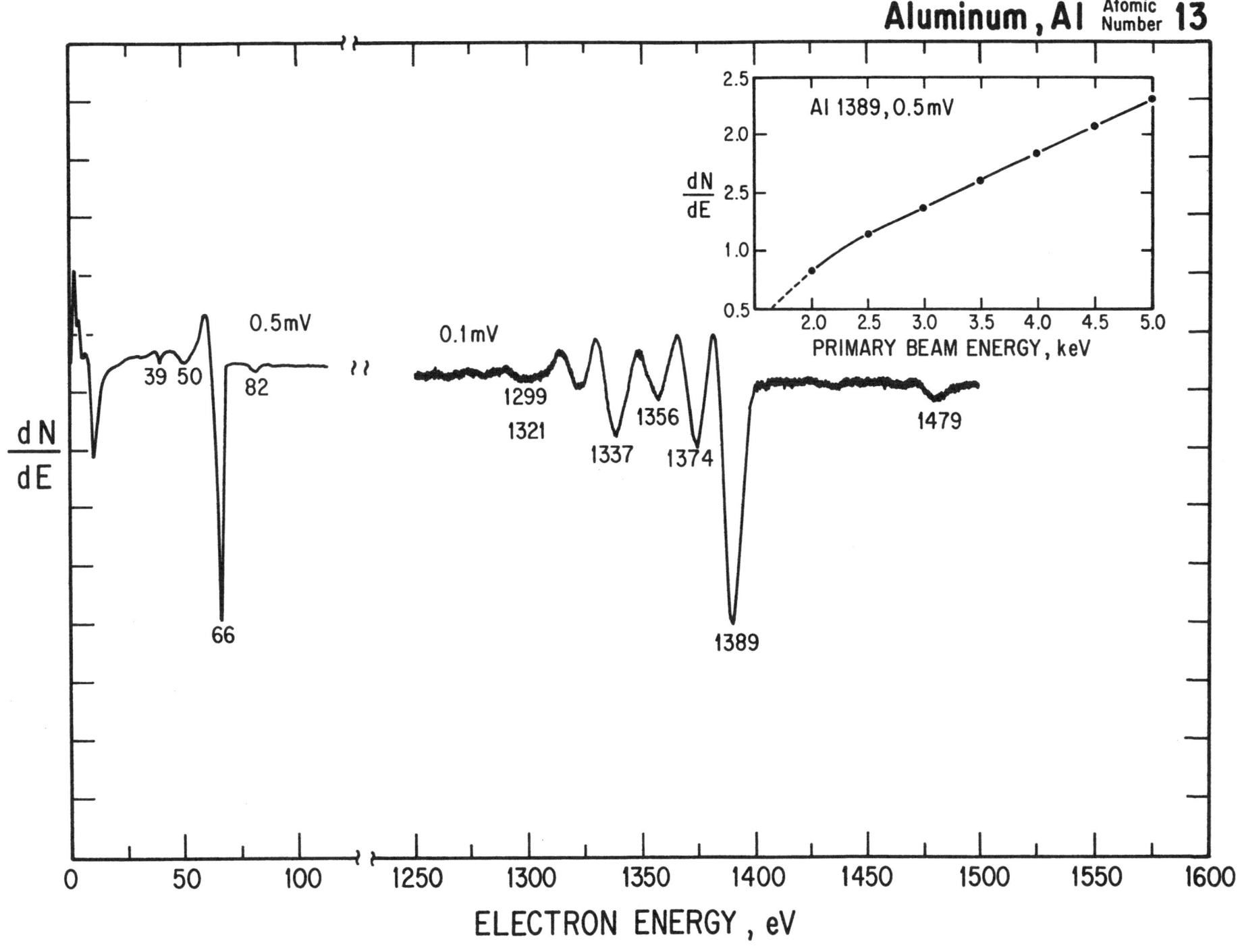

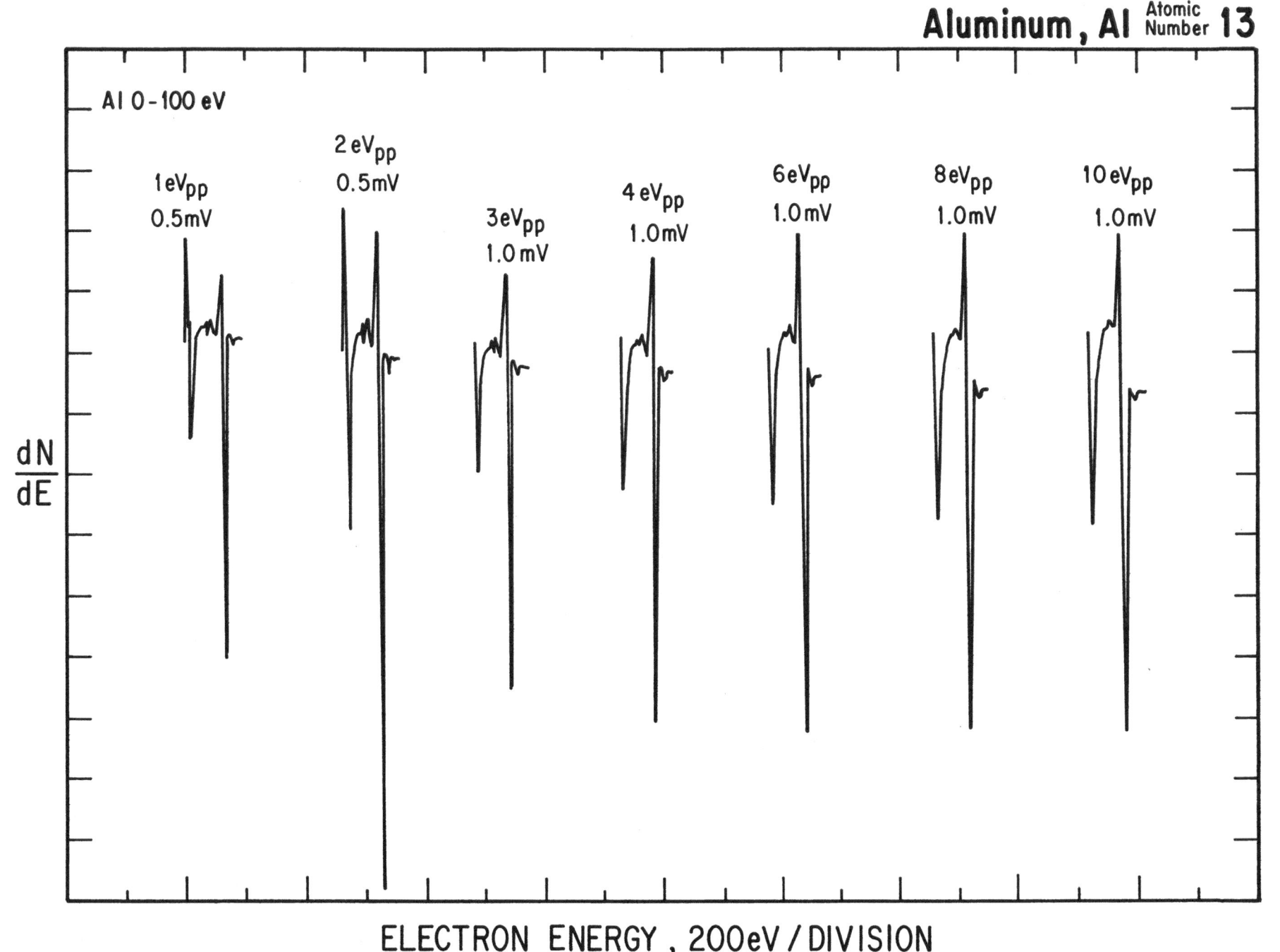

Al 0-100 eV

1 eV$_{pp}$
0.5 mV

2 eV$_{pp}$
0.5 mV

3 eV$_{pp}$
1.0 mV

4 eV$_{pp}$
1.0 mV

6 eV$_{pp}$
1.0 mV

8 eV$_{pp}$
1.0 mV

10 eV$_{pp}$
1.0 mV

$\dfrac{dN}{dE}$

ELECTRON ENERGY, 200eV / DIVISION

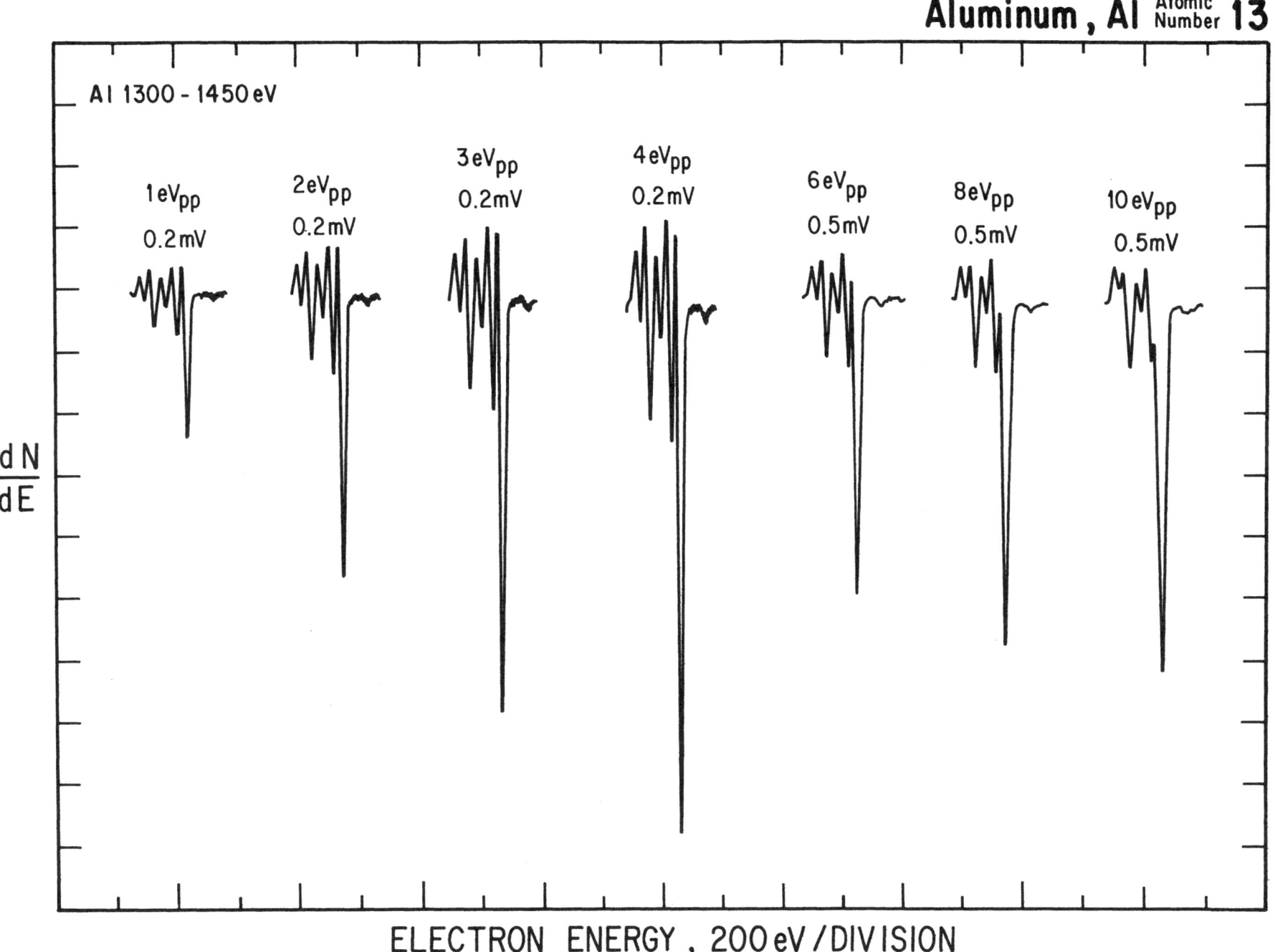

Al 1300 - 1450 eV

1 eV_pp
0.2 mV

2 eV_pp
0.2 mV

3 eV_pp
0.2 mV

4 eV_pp
0.2 mV

6 eV_pp
0.5 mV

8 eV_pp
0.5 mV

10 eV_pp
0.5 mV

$\dfrac{dN}{dE}$

ELECTRON ENERGY , 200 eV / DIVISION

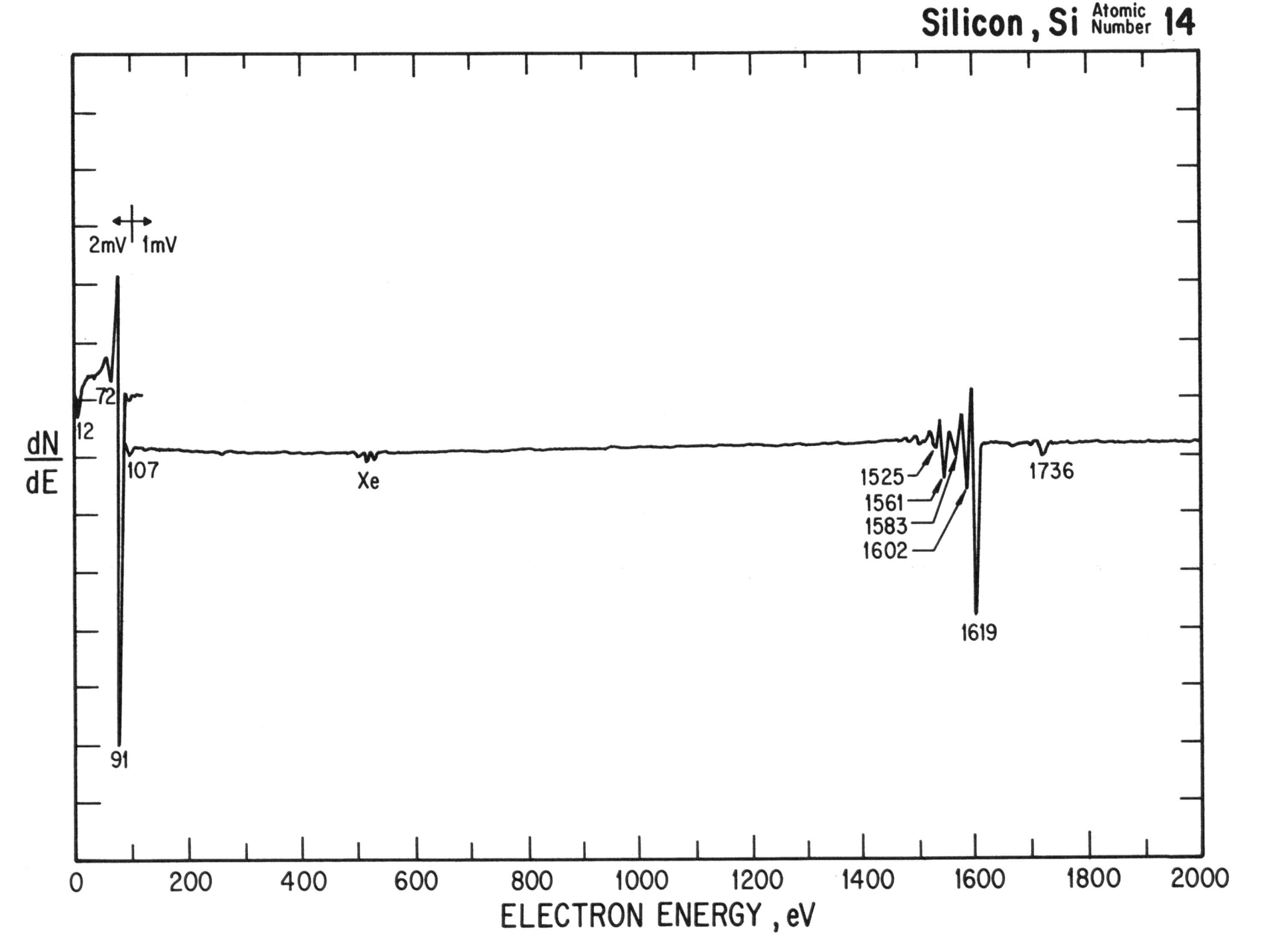

Silicon, Si Atomic Number **14**

2mV 1mV

$\dfrac{dN}{dE}$

12
72
107
91
Xe
1525
1561
1583
1602
1619
1736

ELECTRON ENERGY, eV

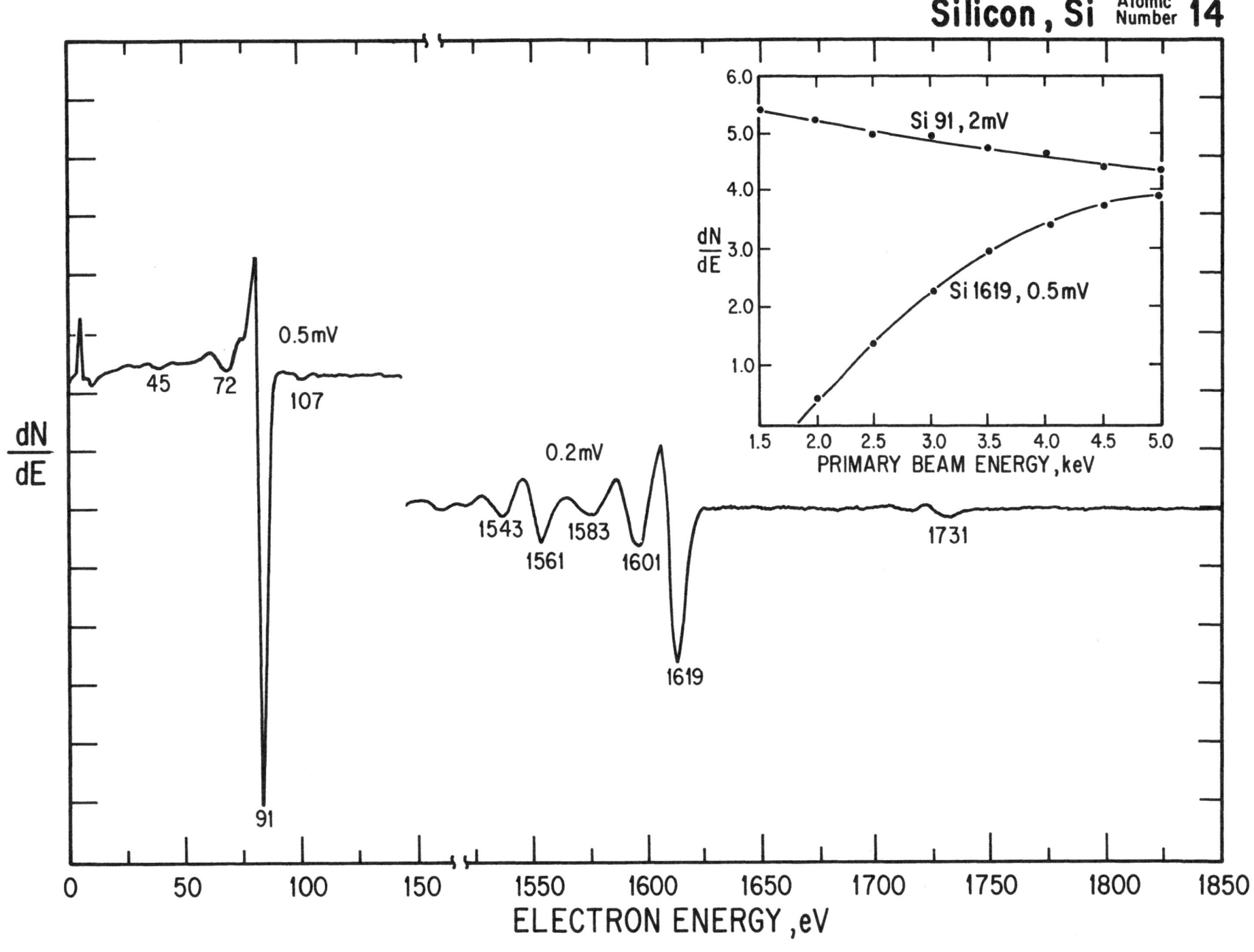

Silicon, Si   Atomic Number **14**

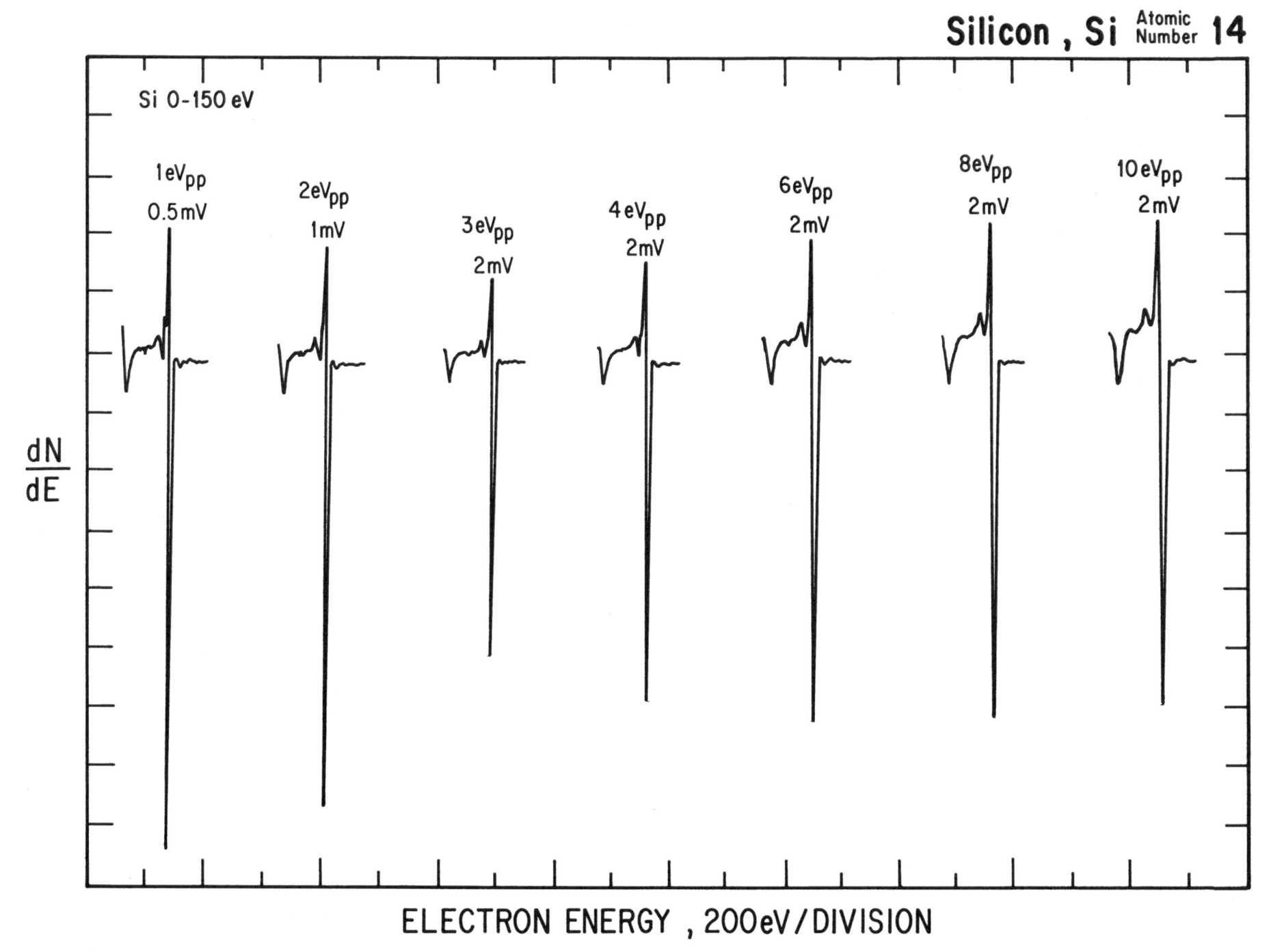

ELECTRON ENERGY , 200eV/DIVISION

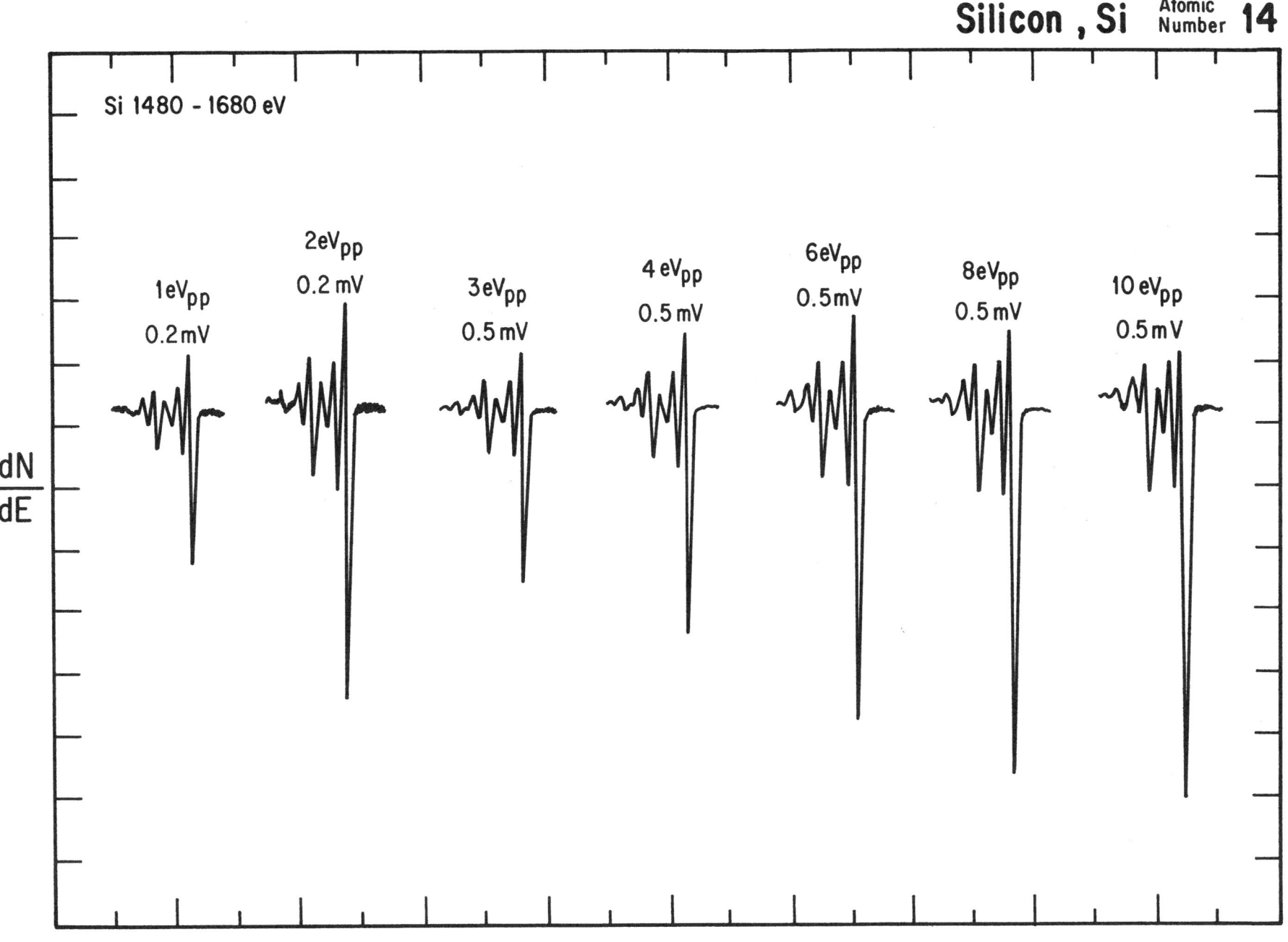

Si 1480 - 1680 eV

1eV$_{pp}$ 0.2 mV

2eV$_{pp}$ 0.2 mV

3eV$_{pp}$ 0.5 mV

4 eV$_{pp}$ 0.5 mV

6eV$_{pp}$ 0.5 mV

8eV$_{pp}$ 0.5 mV

10 eV$_{pp}$ 0.5 mV

$\dfrac{dN}{dE}$

ELECTRON ENERGY , 200 eV/DIVISION

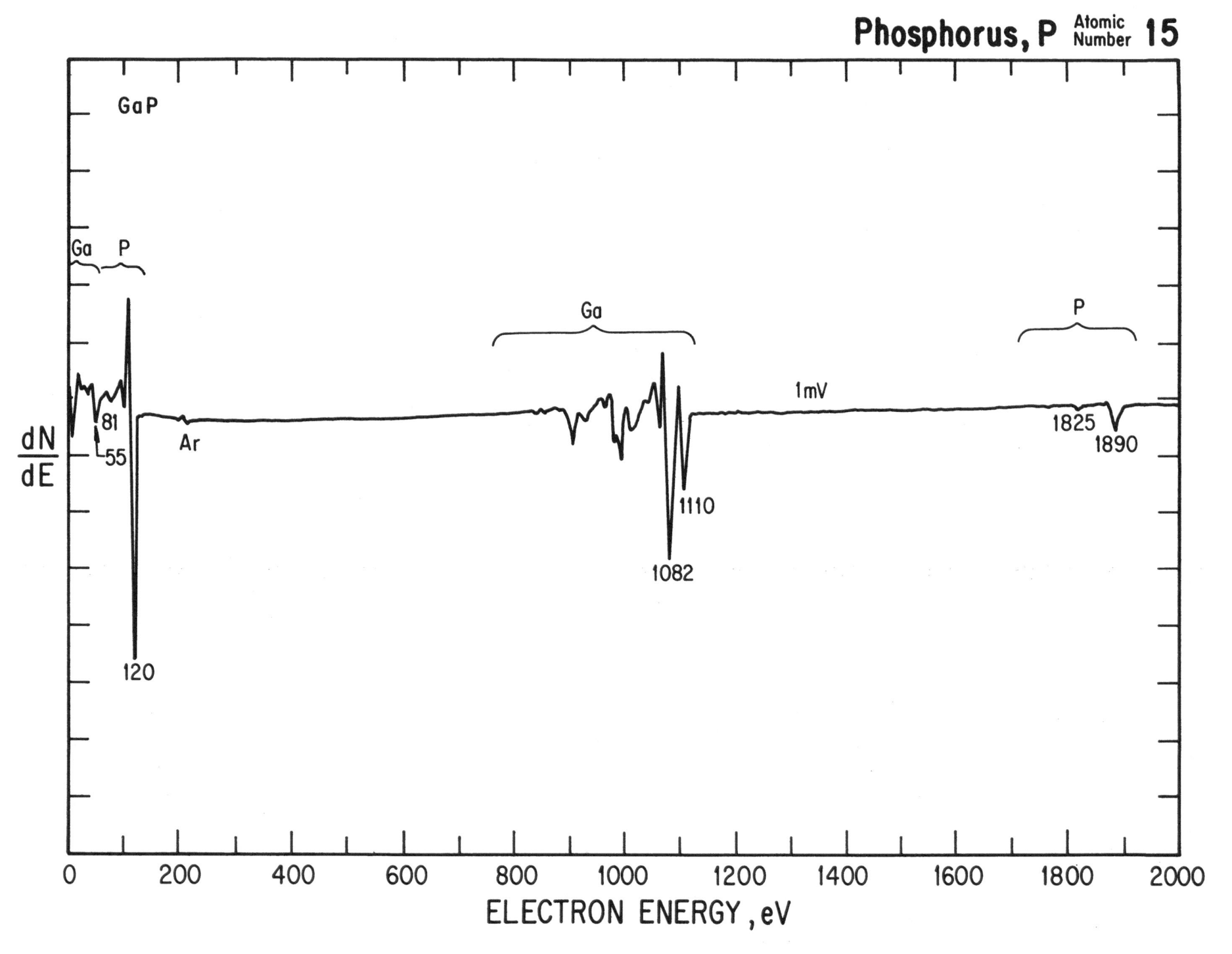

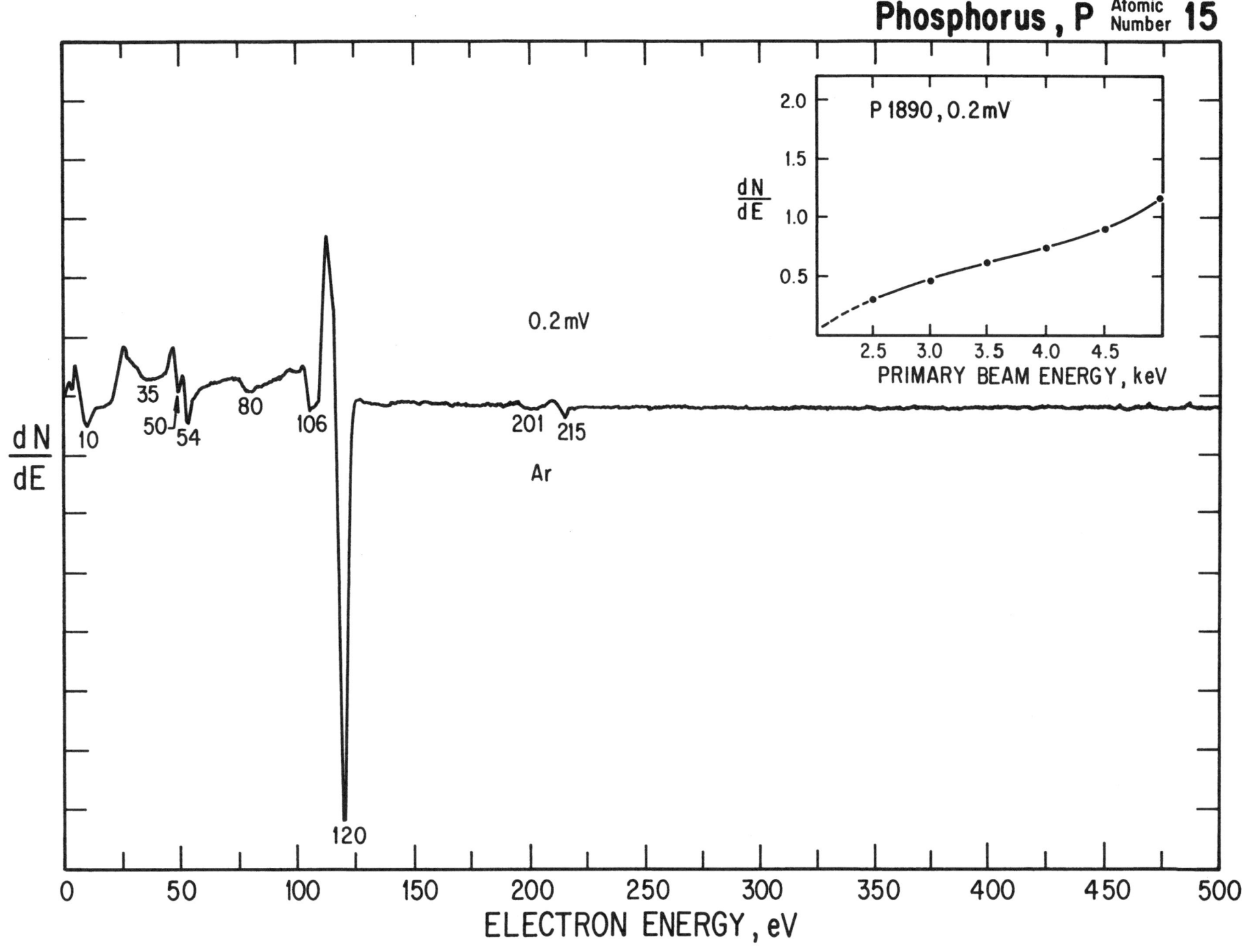

Phosphorus, P Atomic Number 15

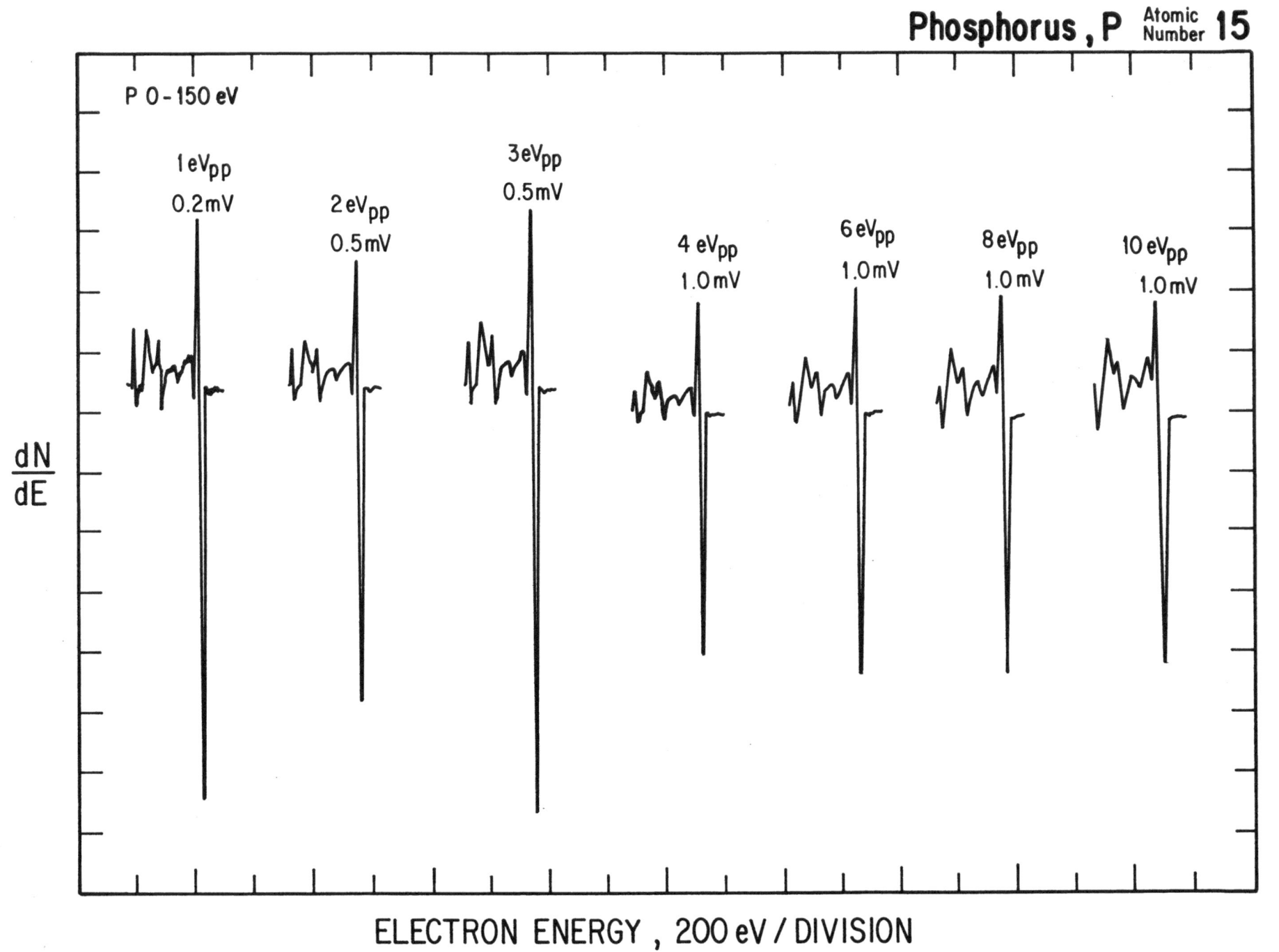

Phosphorus , P Atomic Number **15**

P 0-150 eV

1 eV$_{pp}$
0.2 mV

2 eV$_{pp}$
0.5 mV

3 eV$_{pp}$
0.5 mV

4 eV$_{pp}$
1.0 mV

6 eV$_{pp}$
1.0 mV

8 eV$_{pp}$
1.0 mV

10 eV$_{pp}$
1.0 mV

$\dfrac{dN}{dE}$

ELECTRON ENERGY , 200 eV / DIVISION

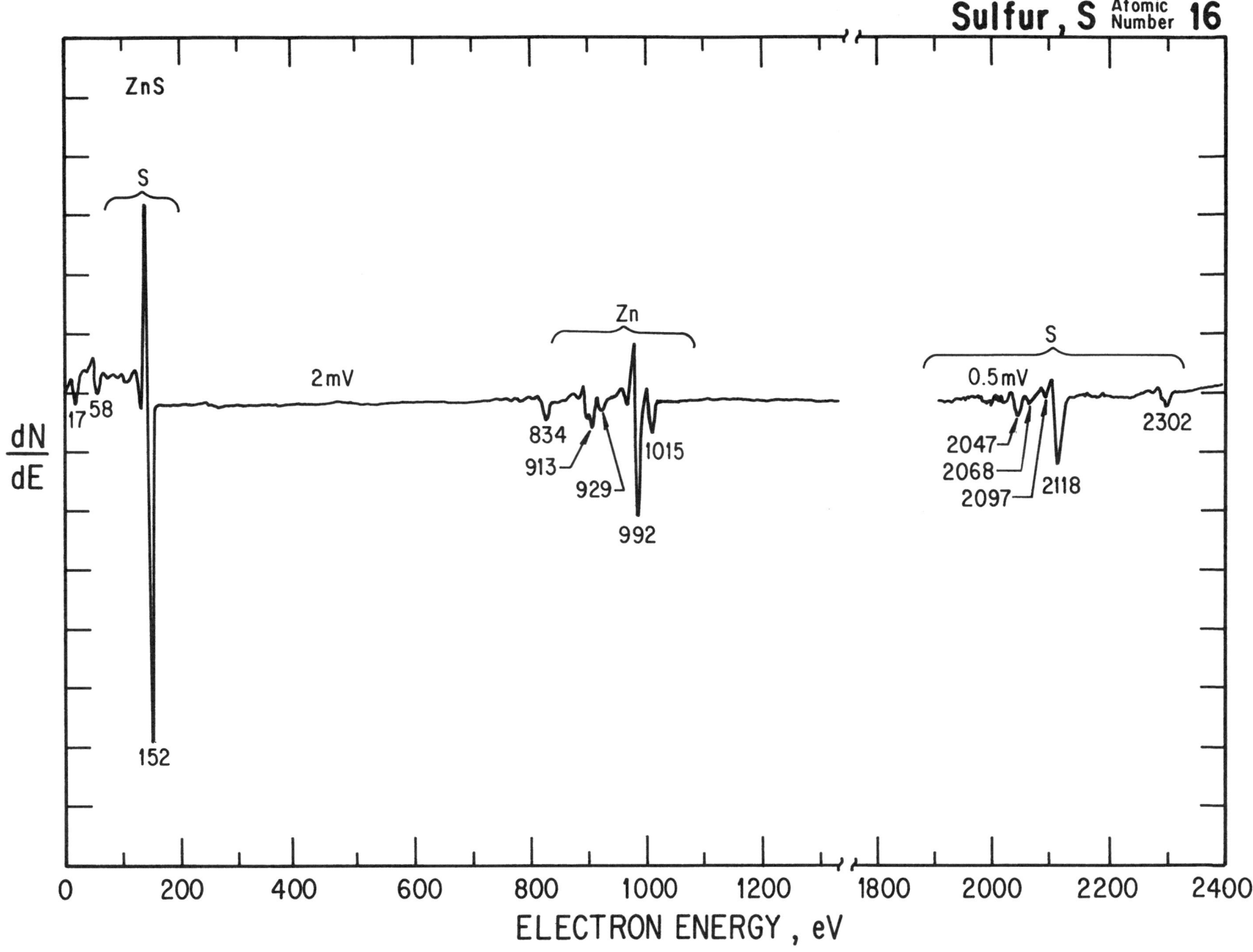

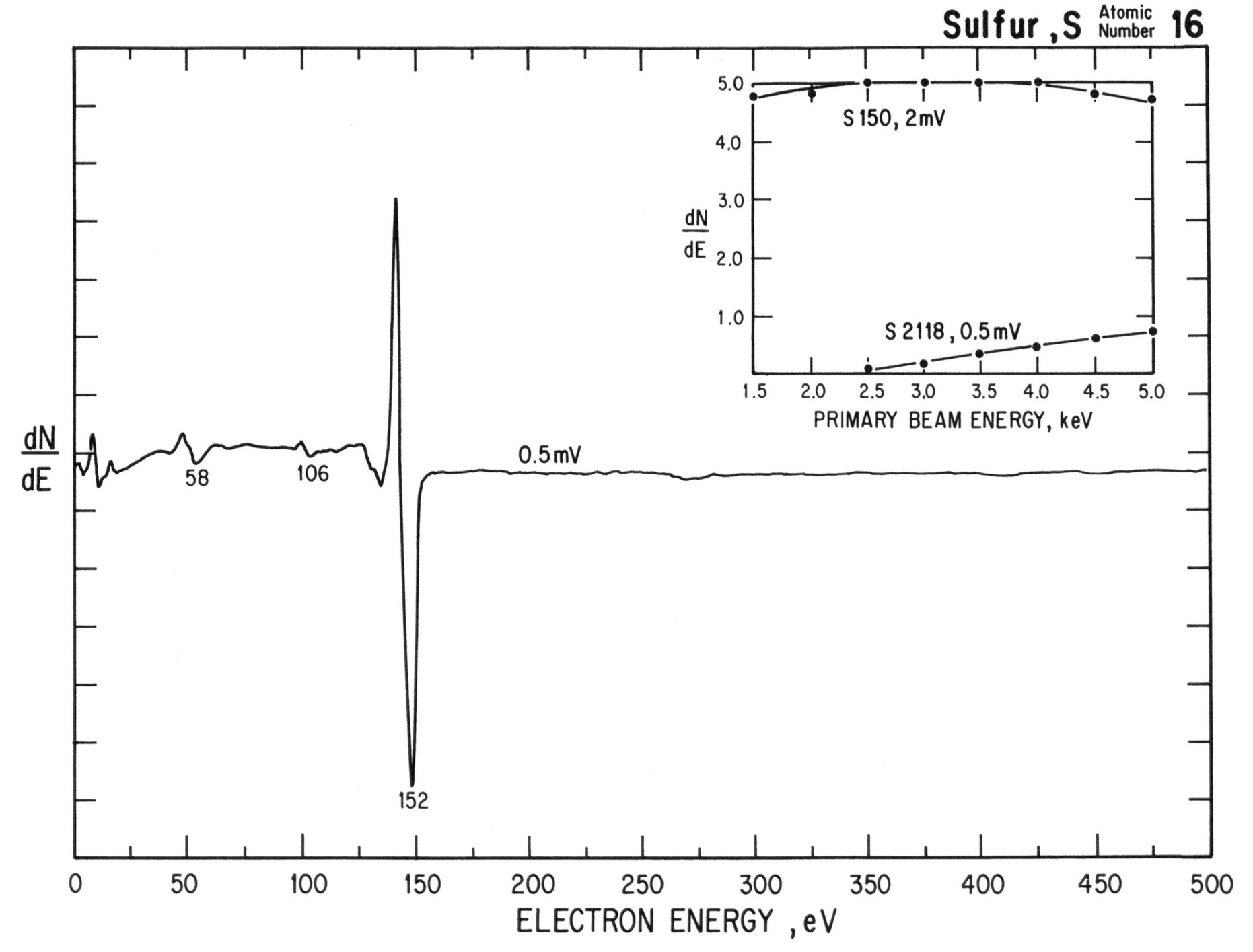

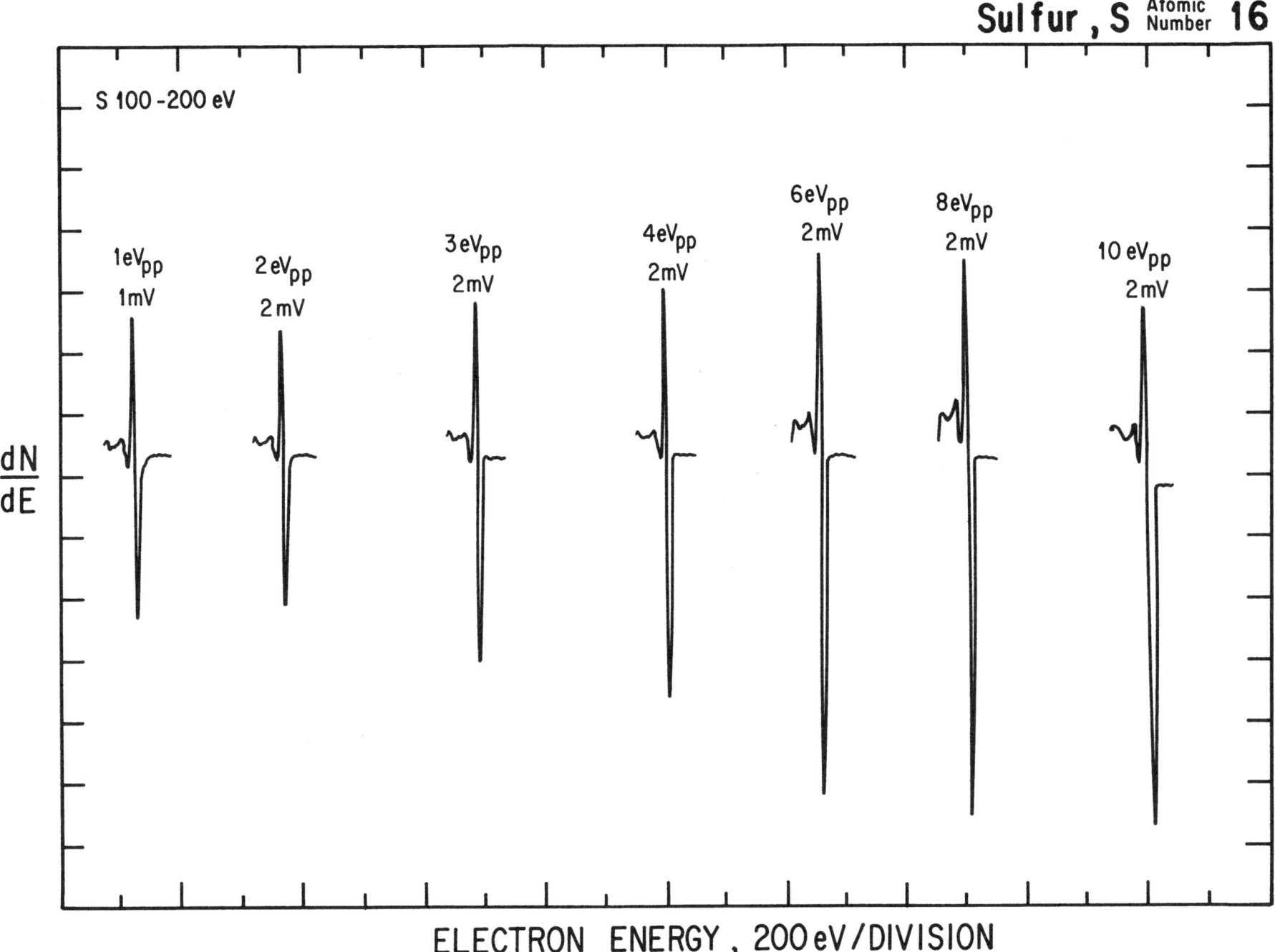

S 100 - 200 eV

1eV$_{pp}$
1mV

2eV$_{pp}$
2mV

3eV$_{pp}$
2mV

4eV$_{pp}$
2mV

6eV$_{pp}$
2mV

8eV$_{pp}$
2mV

10 eV$_{pp}$
2mV

$\dfrac{dN}{dE}$

ELECTRON ENERGY , 200 eV / DIVISION

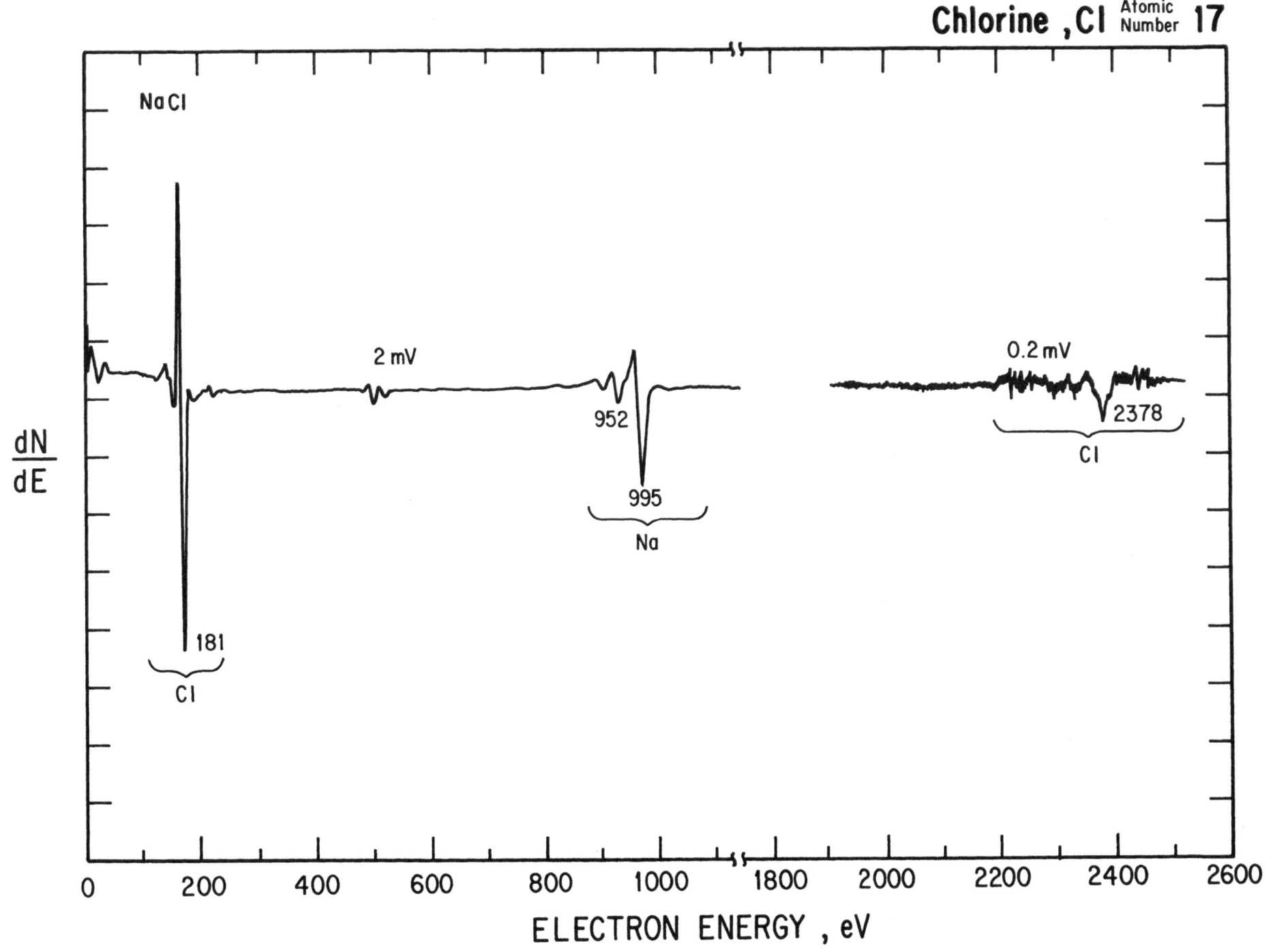

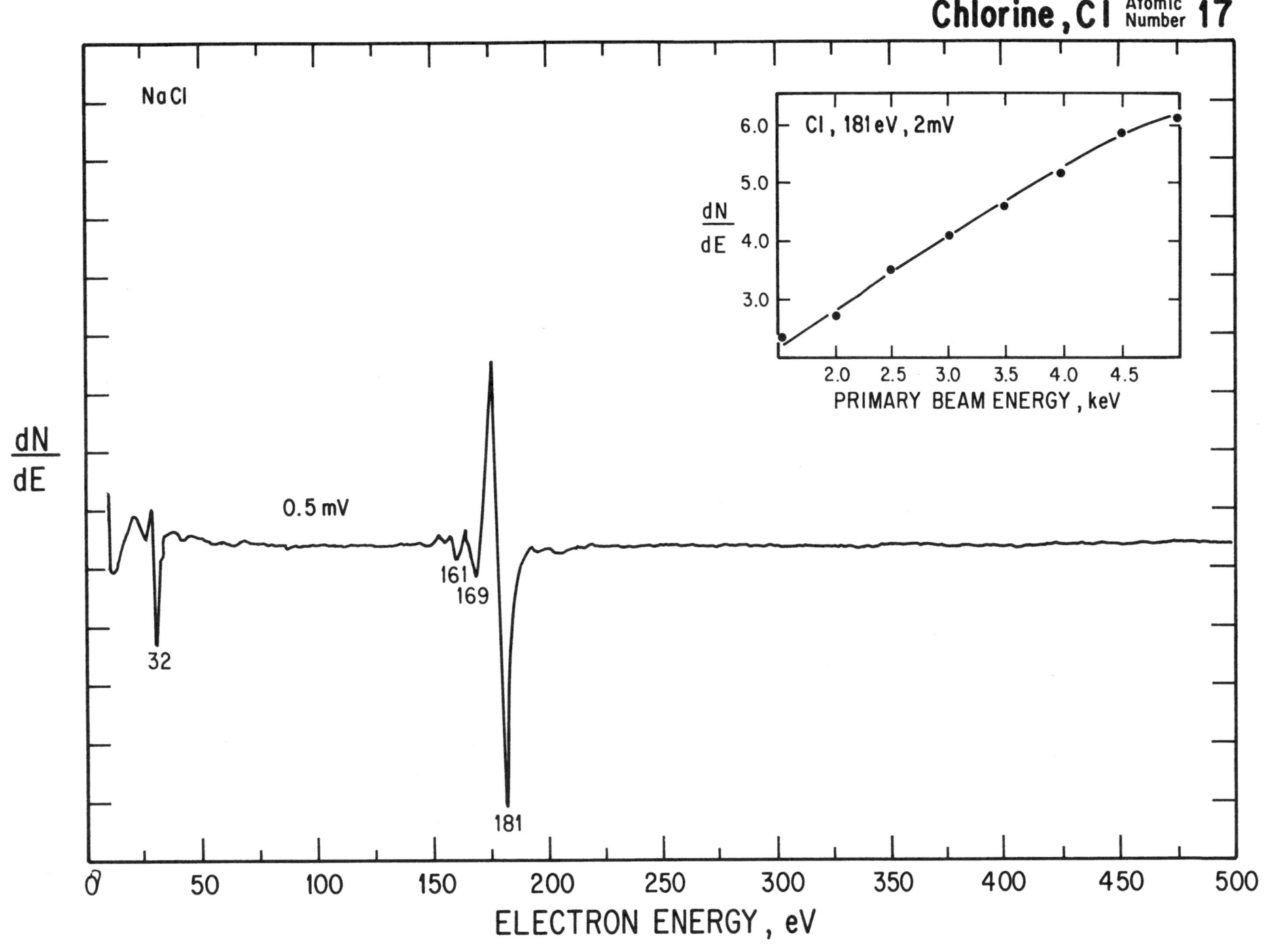

Chlorine, Cl Atomic Number 17

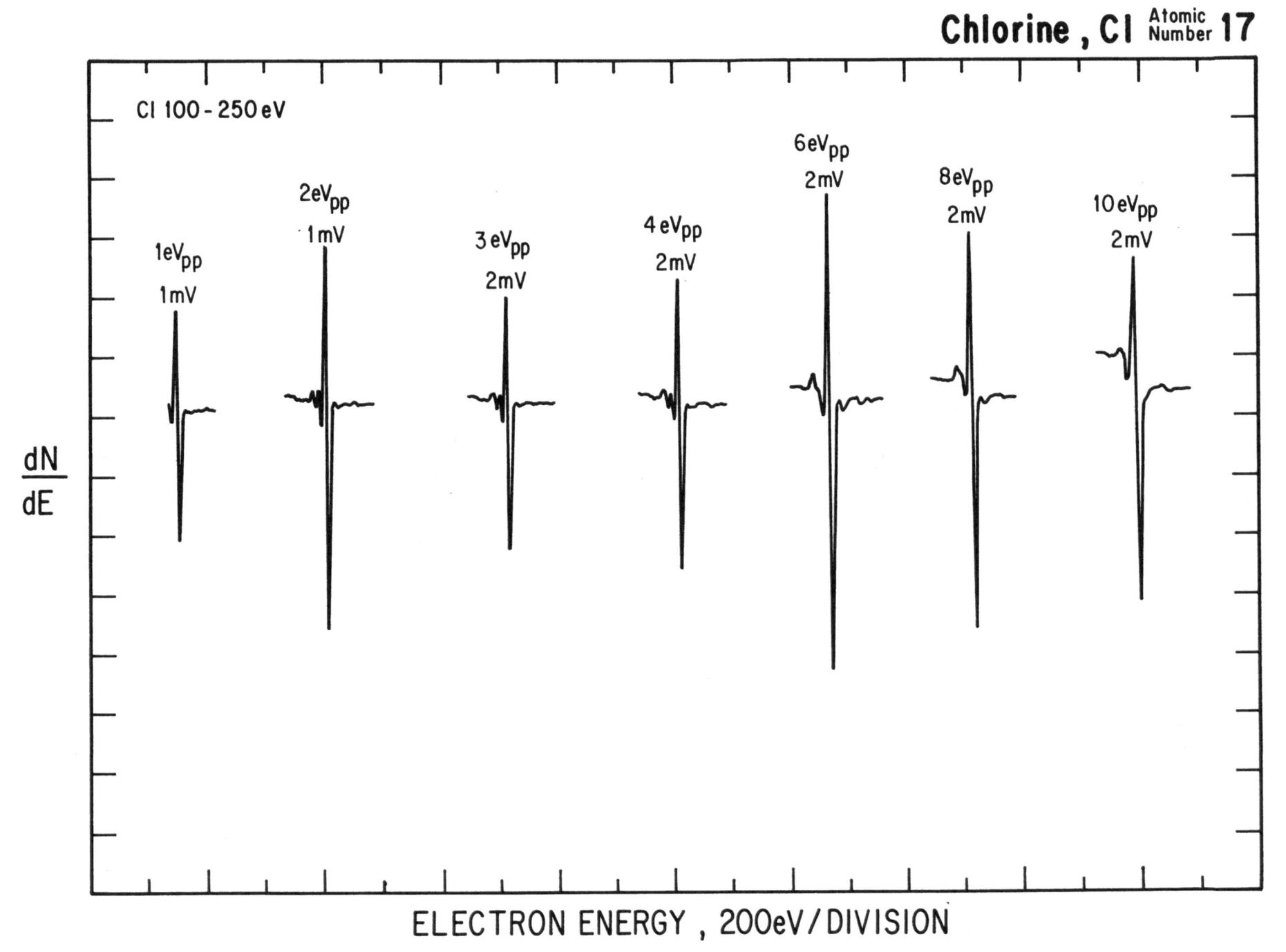

Cl 100 - 250 eV

6eV_pp
2mV

2eV_pp
1mV

8eV_pp
2mV

10eV_pp
2mV

1eV_pp
1mV

3eV_pp
2mV

4eV_pp
2mV

$\dfrac{dN}{dE}$

ELECTRON ENERGY , 200eV/DIVISION

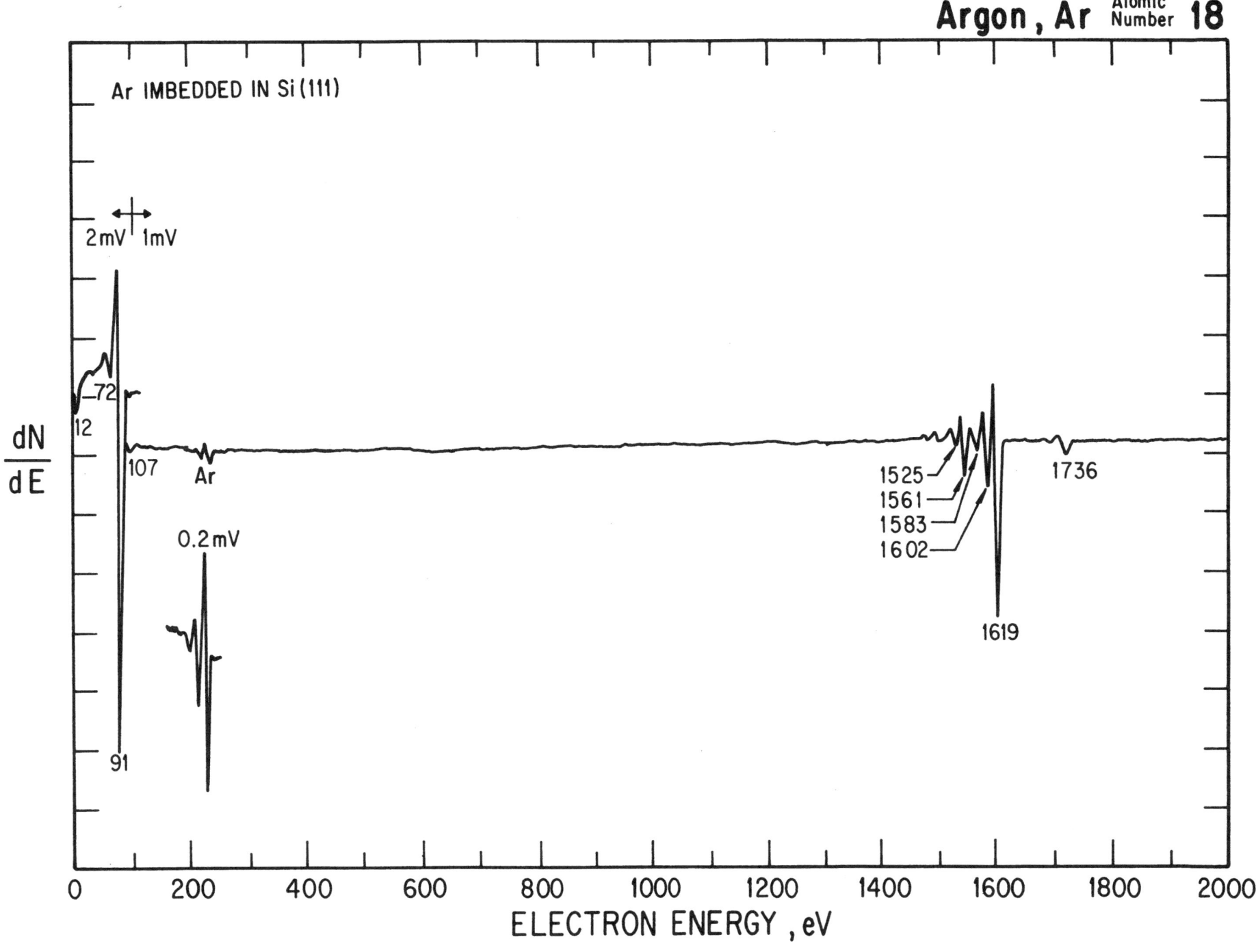

Ar IMBEDDED IN Si (111)

2mV 1mV

$\frac{dN}{dE}$

12
72
107
Ar
91
0.2mV

1525
1561
1583
1602
1619
1736

ELECTRON ENERGY , eV

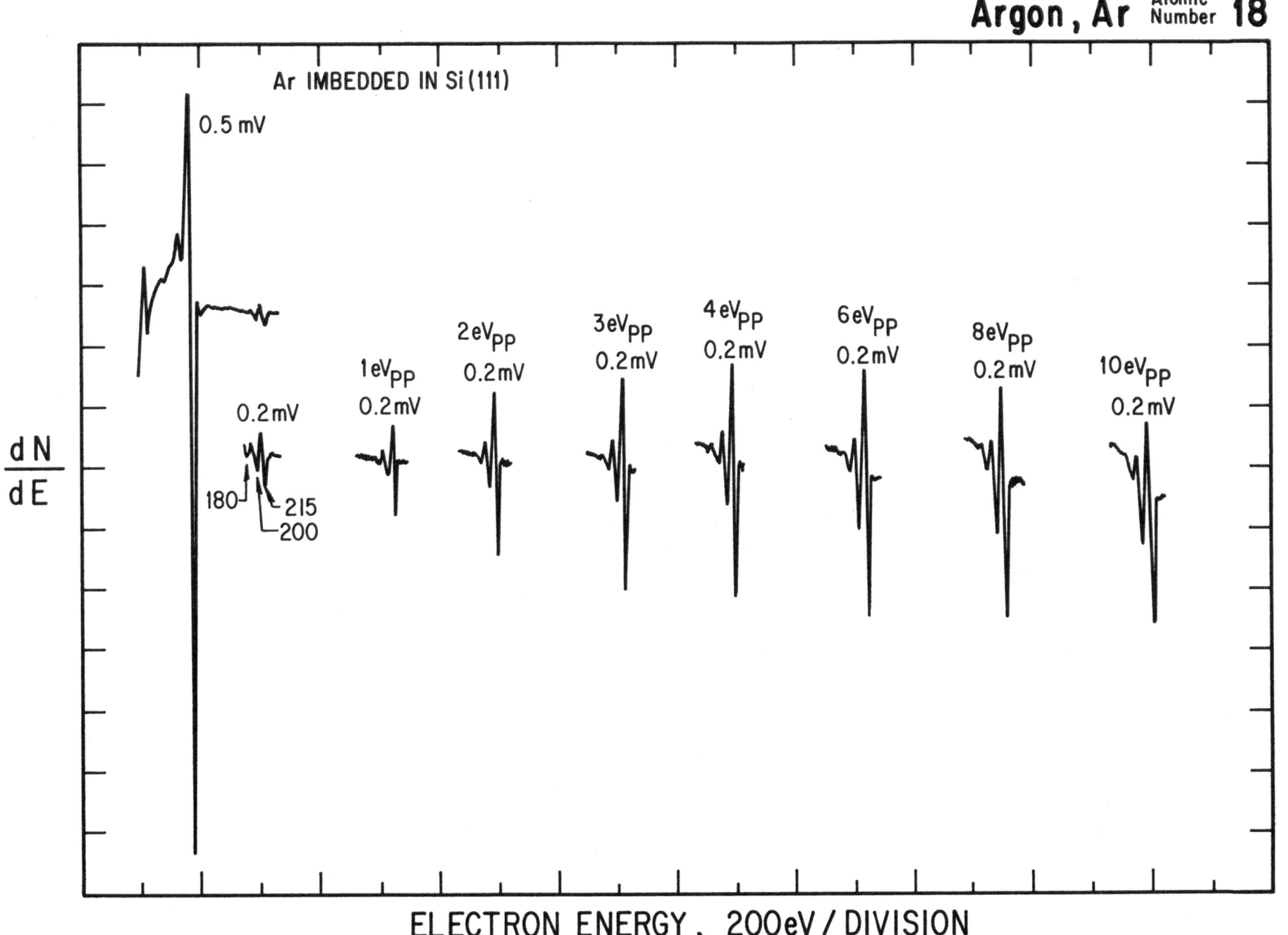

Argon, Ar Atomic Number **18**

Ar IMBEDDED IN Si (111)

0.5 mV

0.2 mV

180 — 215
— 200

1 eV$_{PP}$
0.2 mV

2 eV$_{PP}$
0.2 mV

3 eV$_{PP}$
0.2 mV

4 eV$_{PP}$
0.2 mV

6 eV$_{PP}$
0.2 mV

8 eV$_{PP}$
0.2 mV

10 eV$_{PP}$
0.2 mV

$\dfrac{dN}{dE}$

ELECTRON ENERGY, 200eV / DIVISION

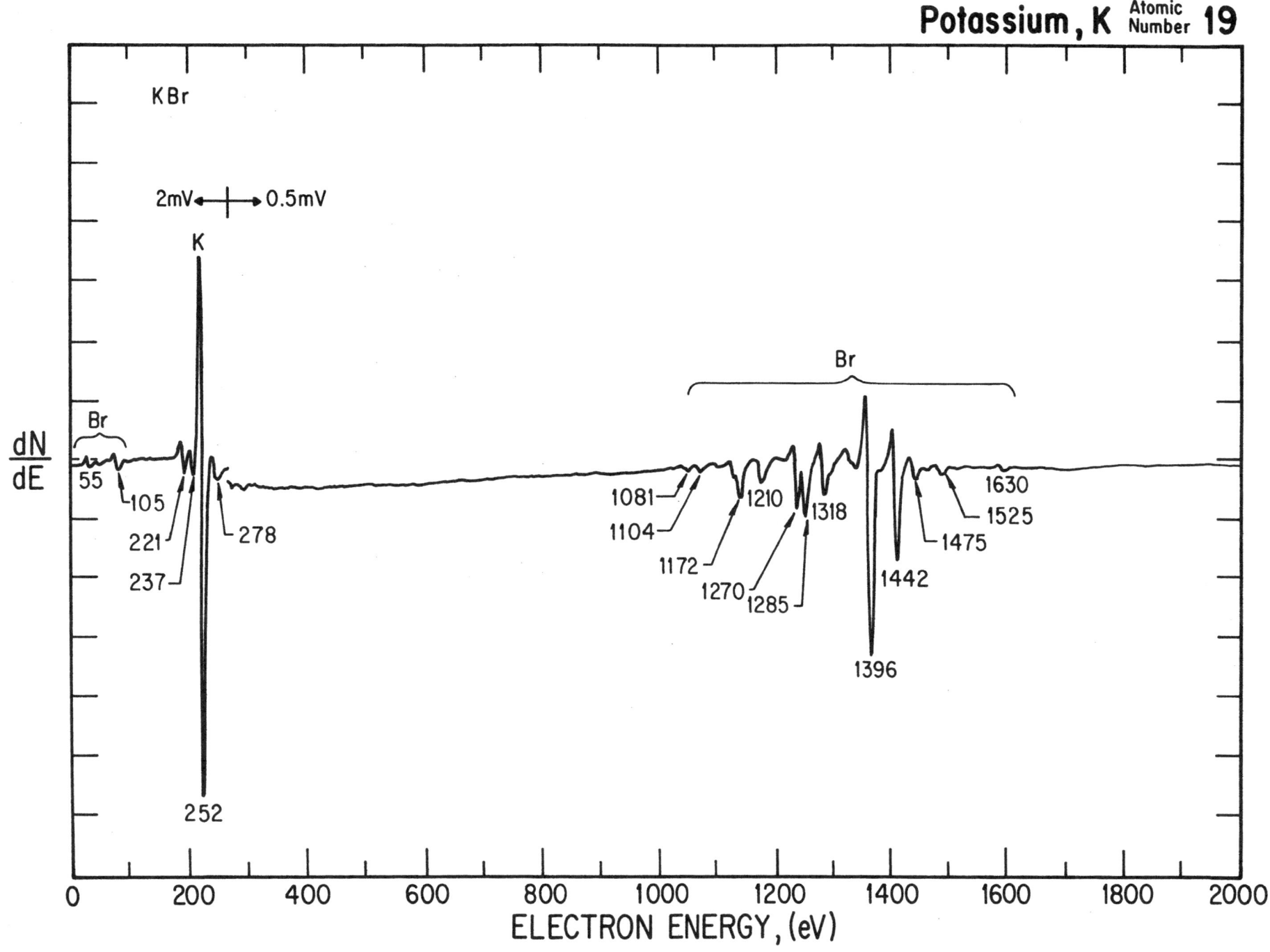

Potassium, K Atomic Number **19**

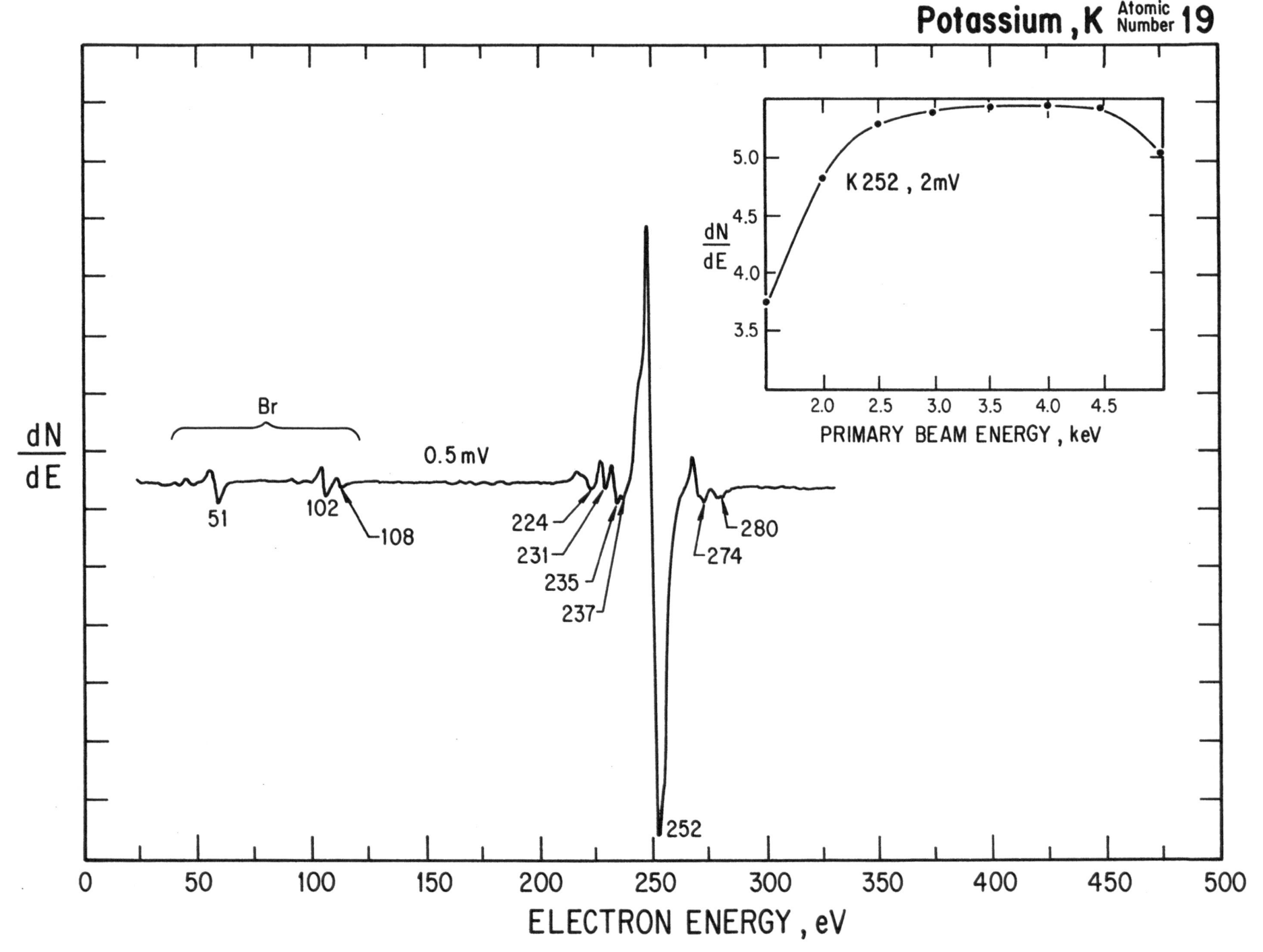

Potassium, K Atomic Number 19

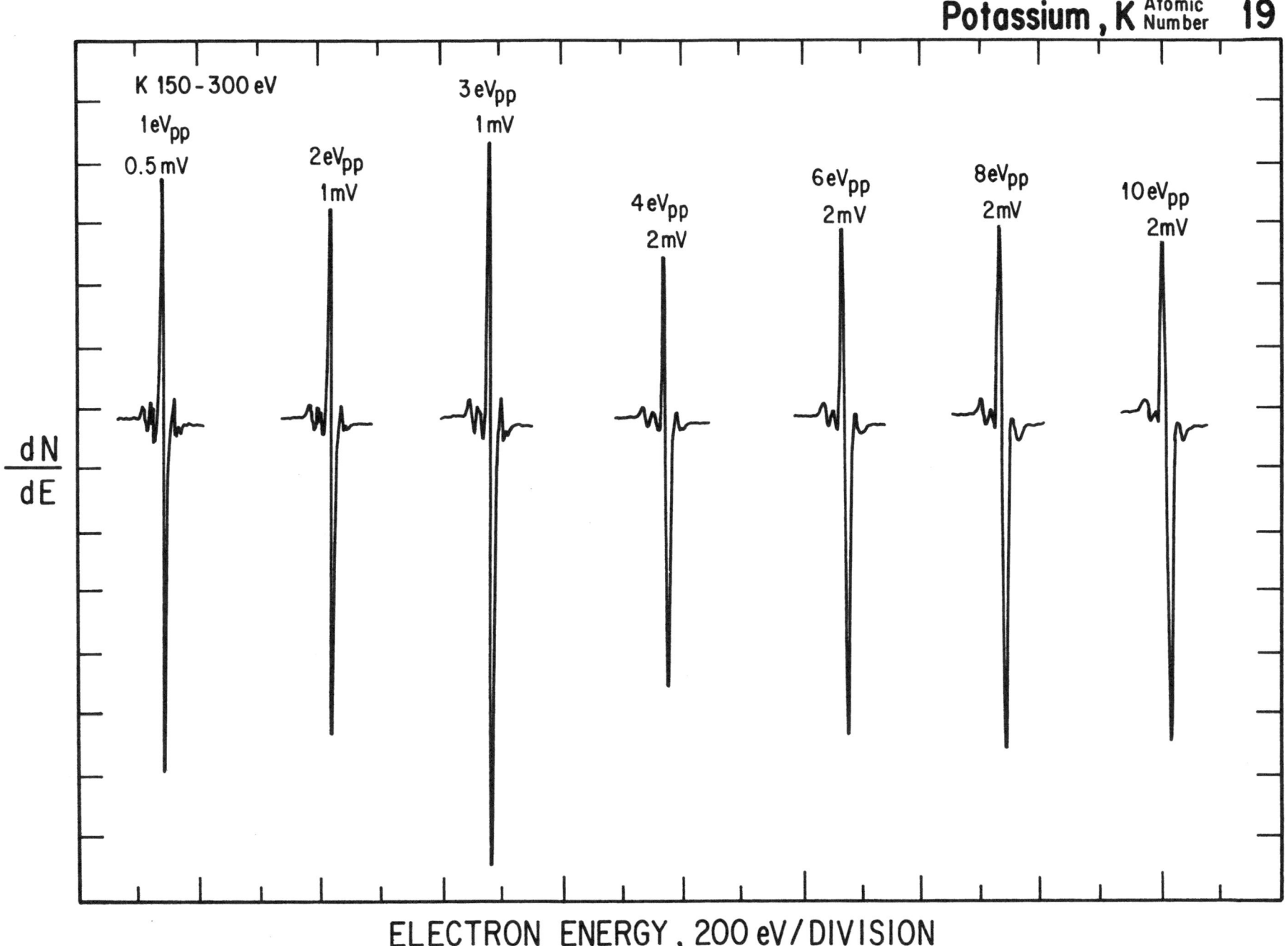

K 150-300 eV

1eV$_{pp}$
0.5 mV

2eV$_{pp}$
1mV

3eV$_{pp}$
1mV

4eV$_{pp}$
2mV

6eV$_{pp}$
2mV

8eV$_{pp}$
2mV

10eV$_{pp}$
2mV

$\dfrac{dN}{dE}$

ELECTRON ENERGY, 200 eV/DIVISION

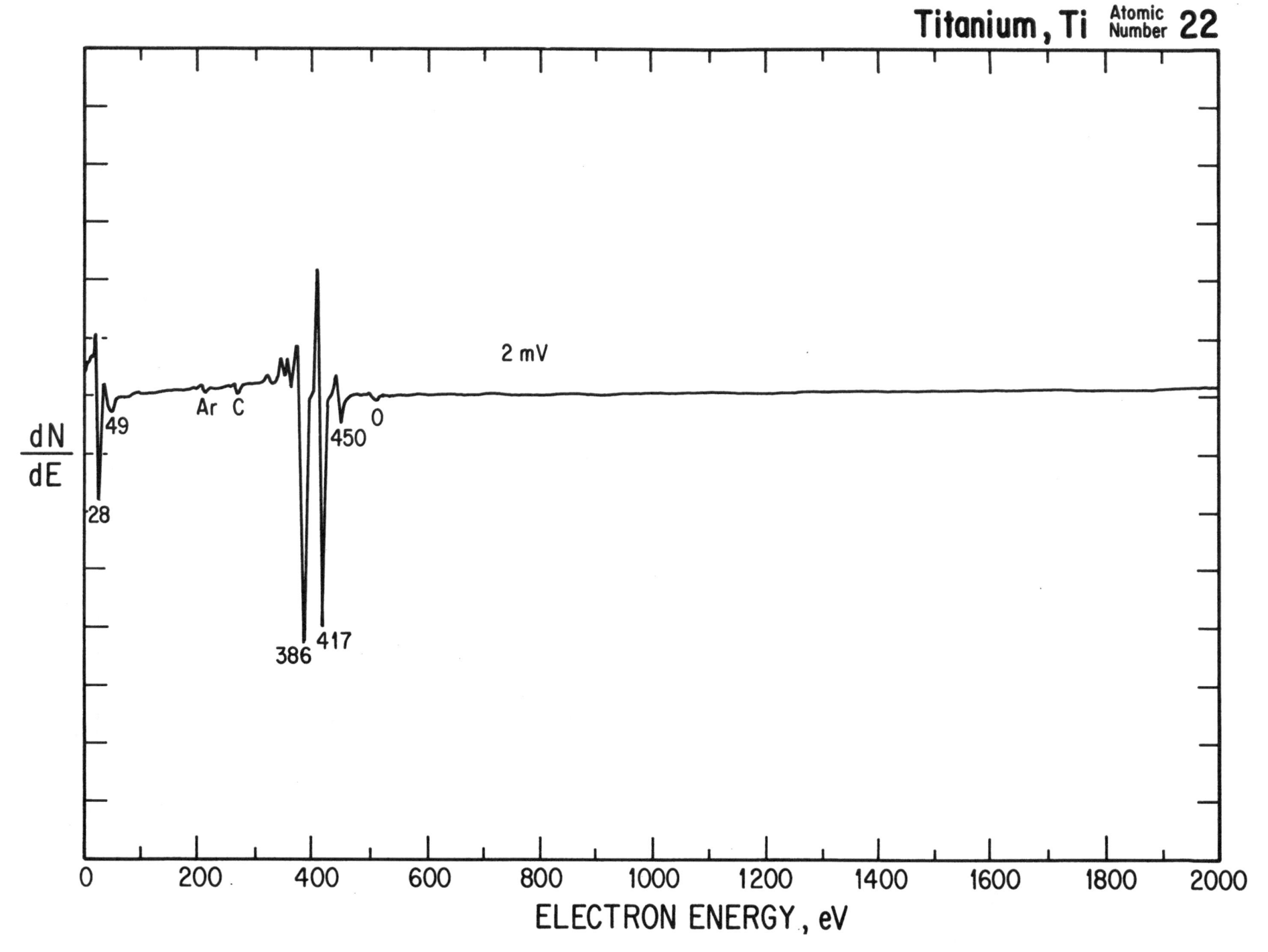

Titanium, Ti Atomic Number 22

$\dfrac{dN}{dE}$

2 mV

Ar C

49

28

386 417

450 0

ELECTRON ENERGY, eV

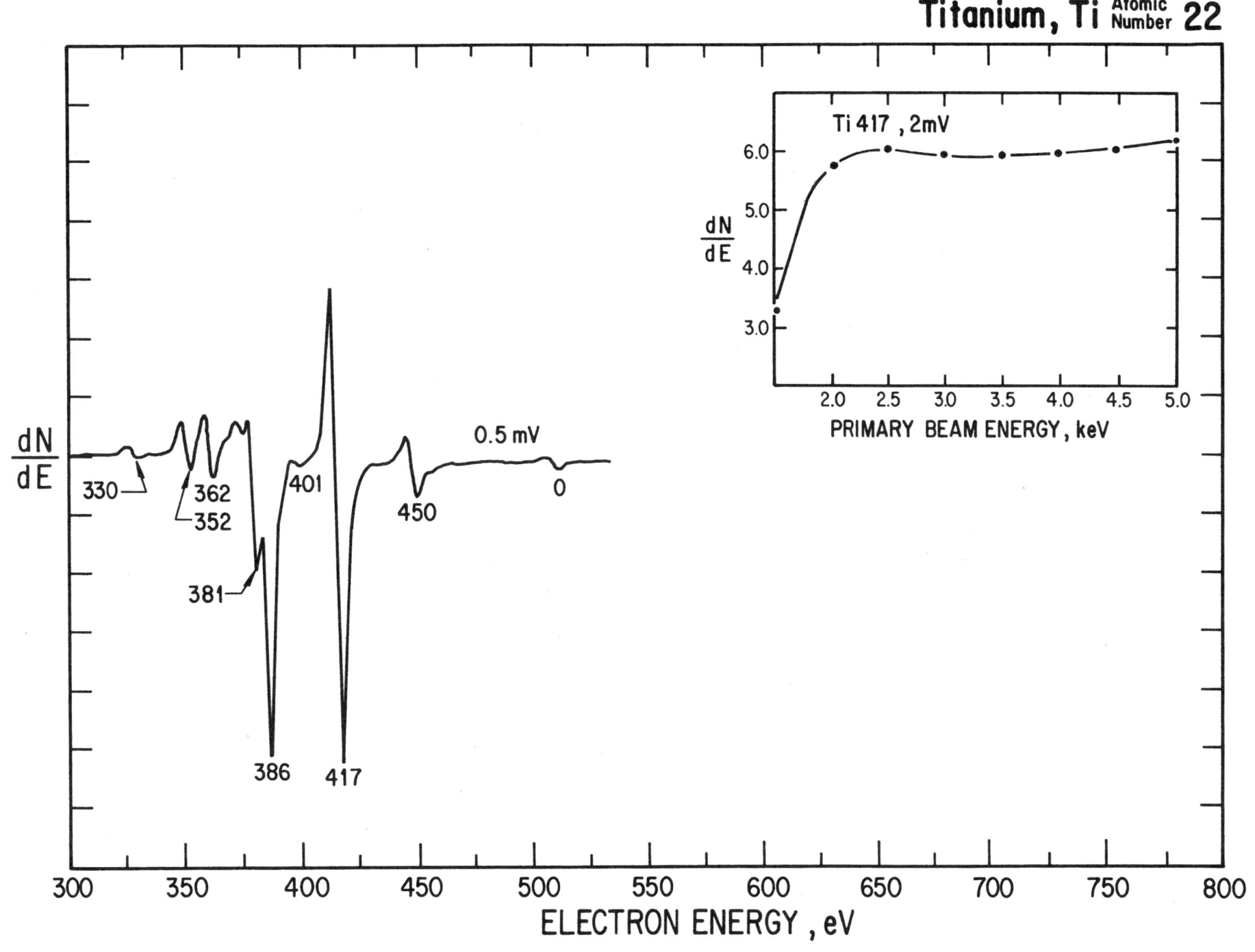

Ti 417 , 2mV

$\frac{dN}{dE}$

6.0

5.0

4.0

3.0

2.0 2.5 3.0 3.5 4.0 4.5 5.0

PRIMARY BEAM ENERGY , keV

$\frac{dN}{dE}$

0.5 mV

330

362

352

381

401

450

0

386

417

ELECTRON ENERGY , eV

Titanium, Ti Atomic Number 22

Ti 350 - 480 eV

1 eV_pp 1 mV

2 eV_pp 2 mV

3 eV_pp 2 mV

4 eV_pp 2 mV

6 eV_pp 2 mV

8 eV_pp 2 mV

10 eV_pp 5 mV

$\dfrac{dN}{dE}$

ELECTRON ENERGY, 200 eV/DIVISION

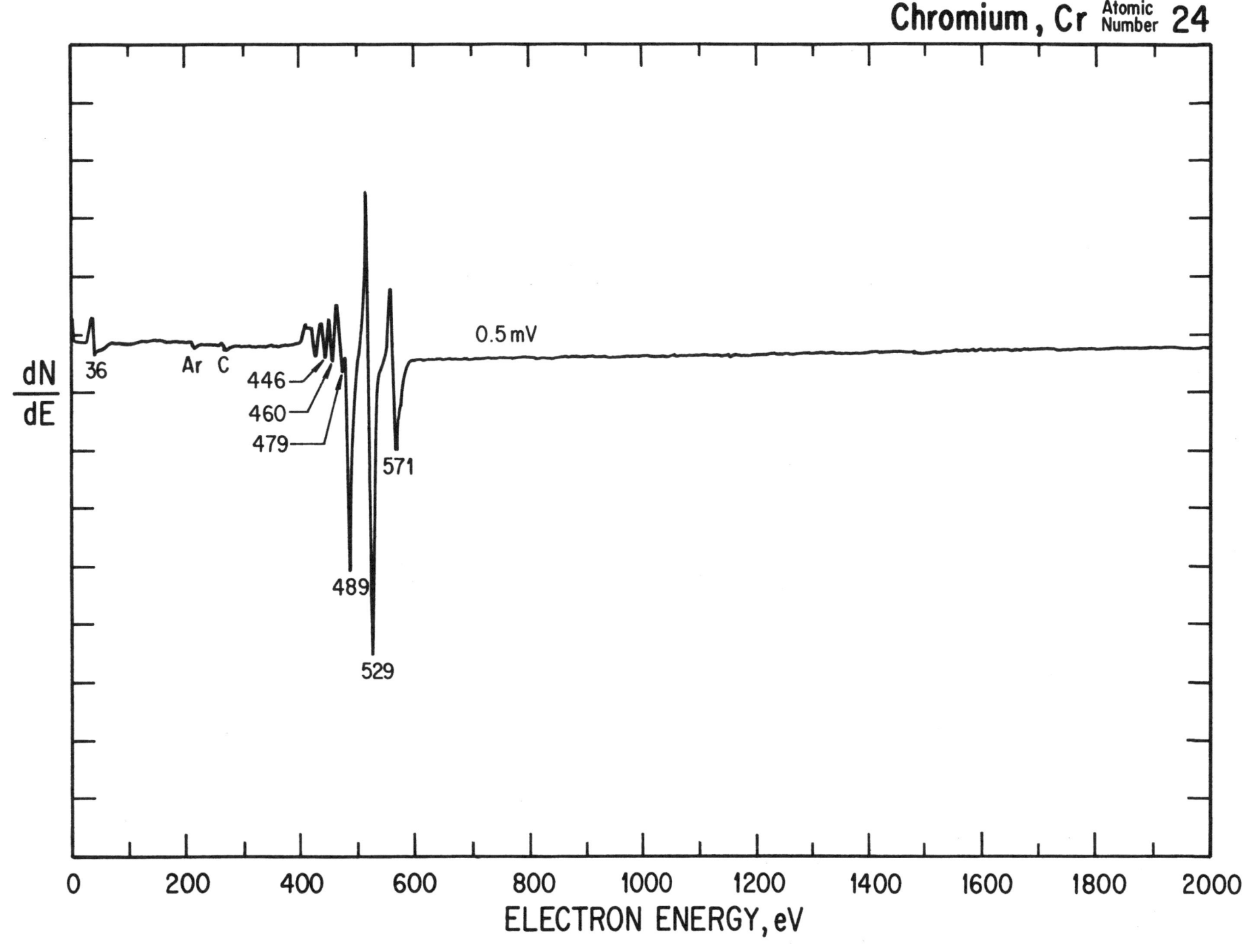

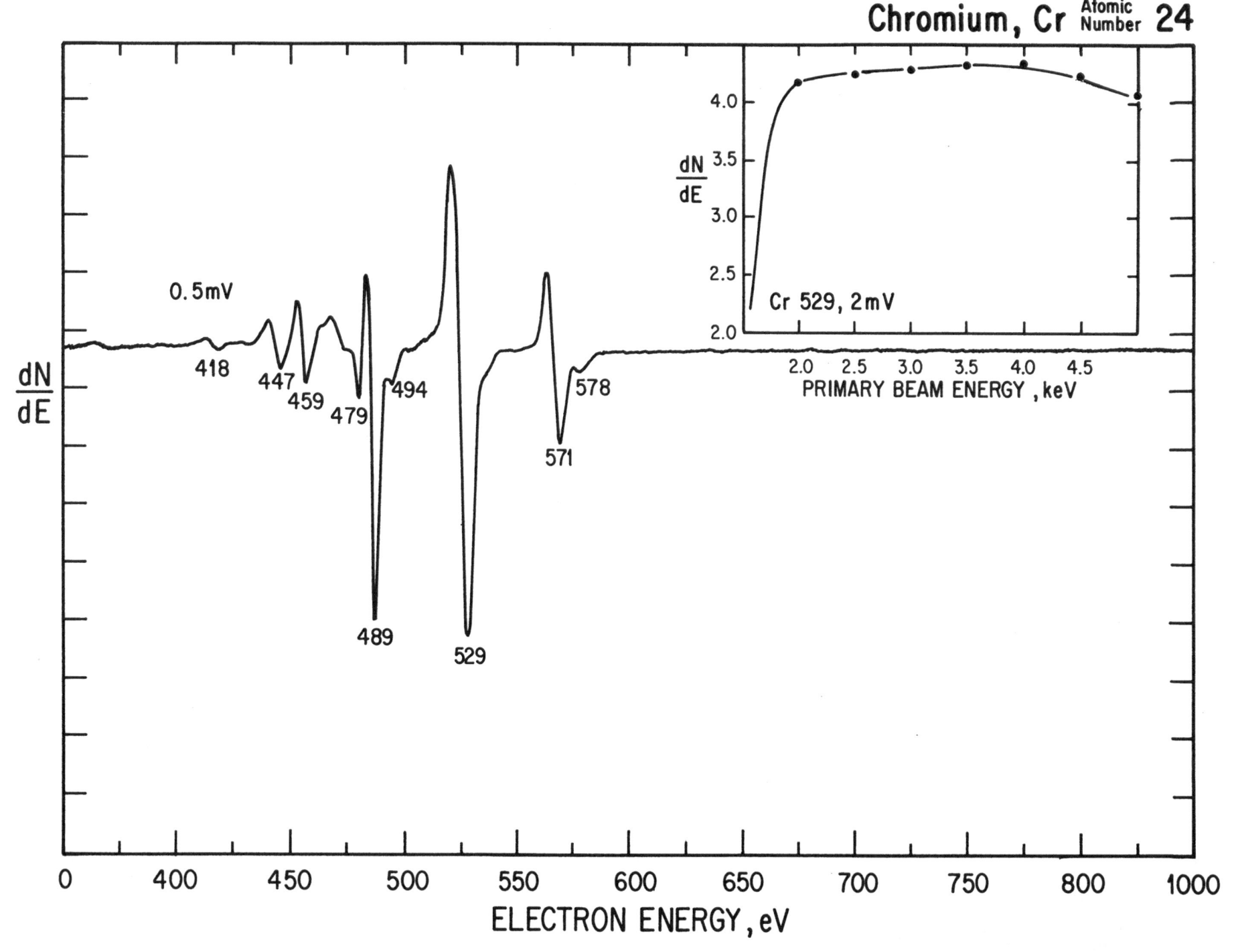

Chromium, Cr Atomic Number 24

$\frac{dN}{dE}$

0.5mV

418
447
459
479
494
489
529
571
578

$\frac{dN}{dE}$

4.0
3.5
3.0
2.5
2.0

Cr 529, 2mV

2.0 2.5 3.0 3.5 4.0 4.5
PRIMARY BEAM ENERGY, keV

0   400   450   500   550   600   650   700   750   800   1000
ELECTRON ENERGY, eV

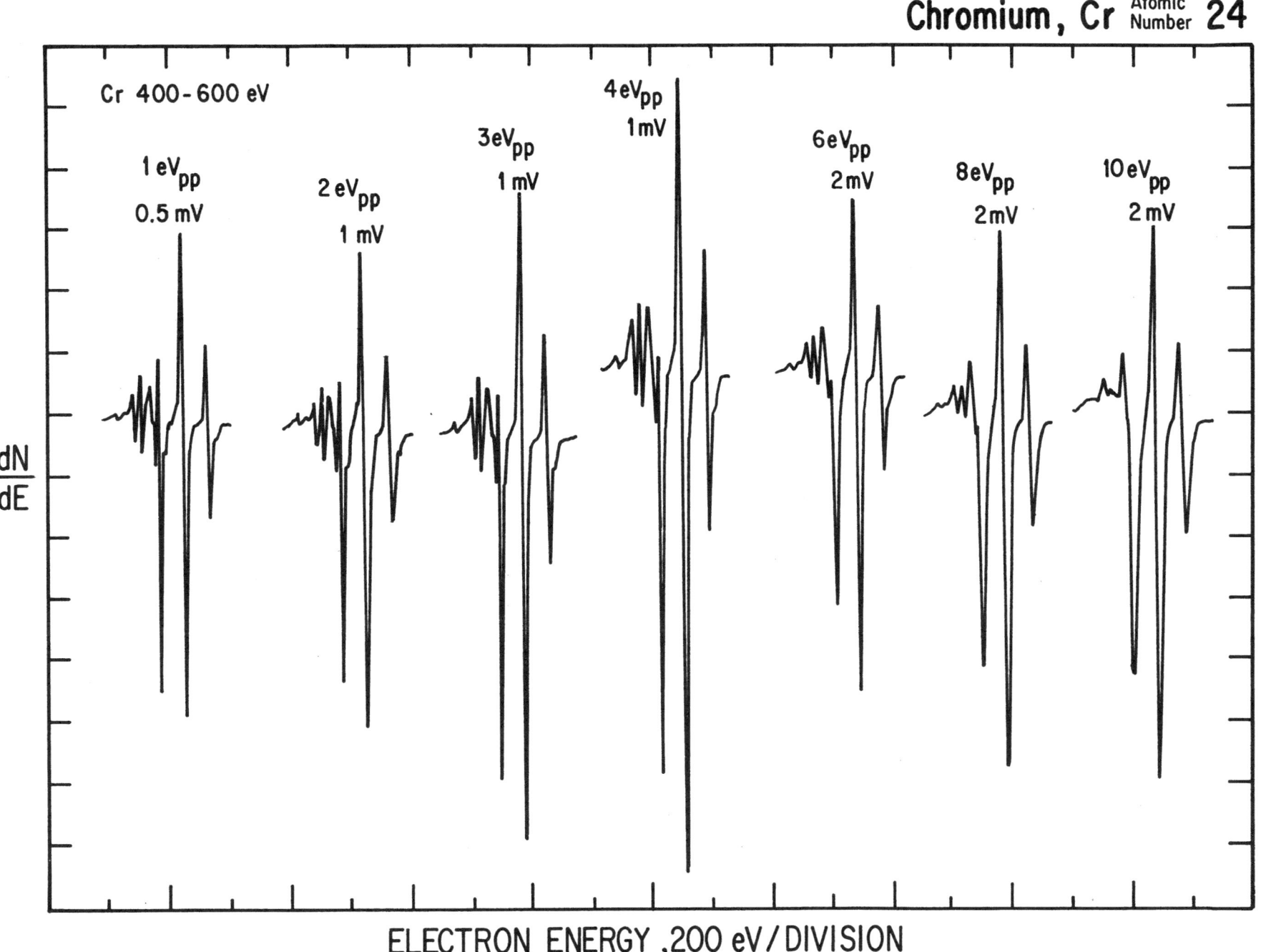

Cr 400-600 eV

1 eV_pp
0.5 mV

2 eV_pp
1 mV

3 eV_pp
1 mV

4 eV_pp
1 mV

6 eV_pp
2 mV

8 eV_pp
2 mV

10 eV_pp
2 mV

$\dfrac{dN}{dE}$

ELECTRON ENERGY ,200 eV/DIVISION

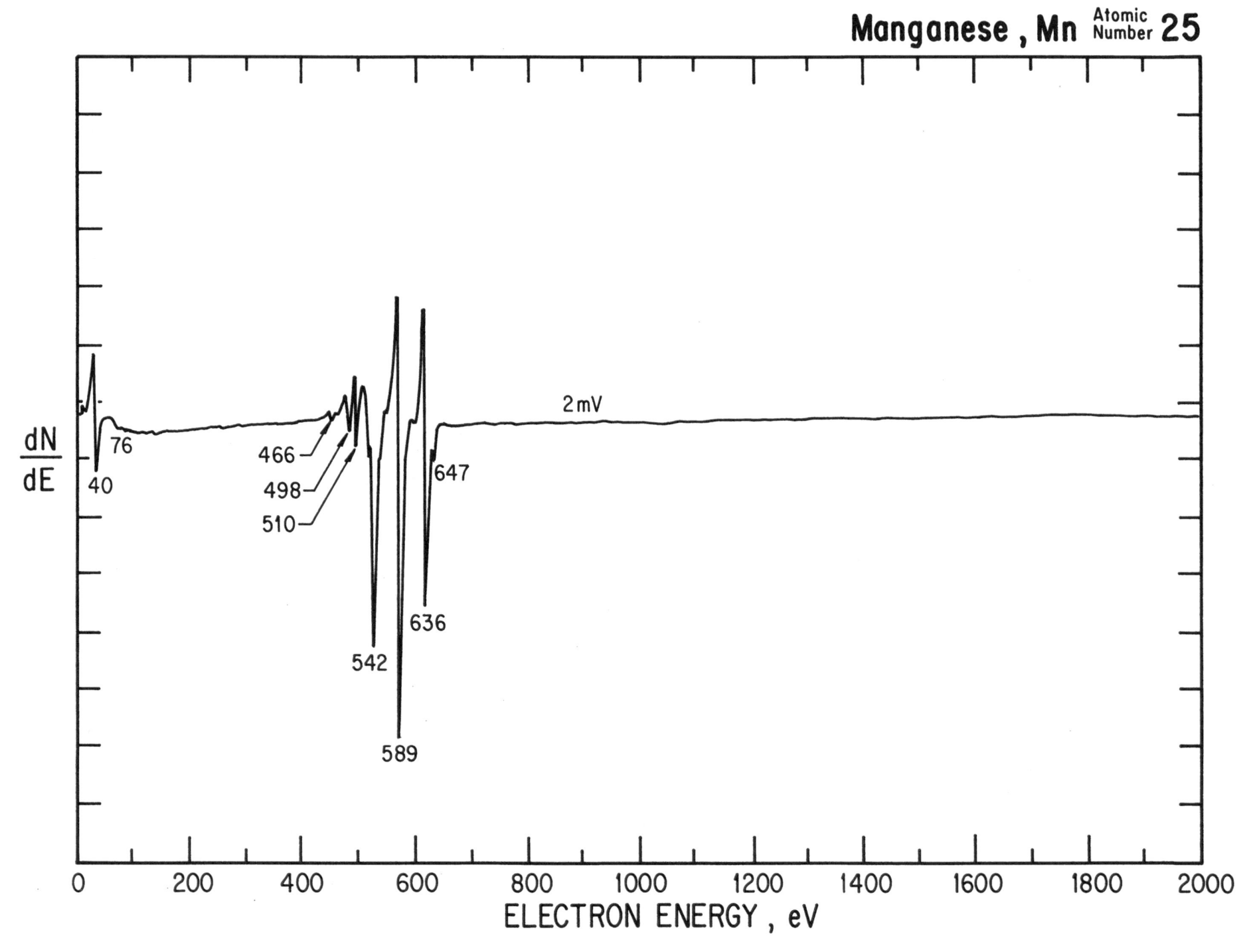

Manganese, Mn Atomic Number **25**

$\dfrac{dN}{dE}$

2 mV

40
76
466
498
510
542
589
636
647

ELECTRON ENERGY, eV

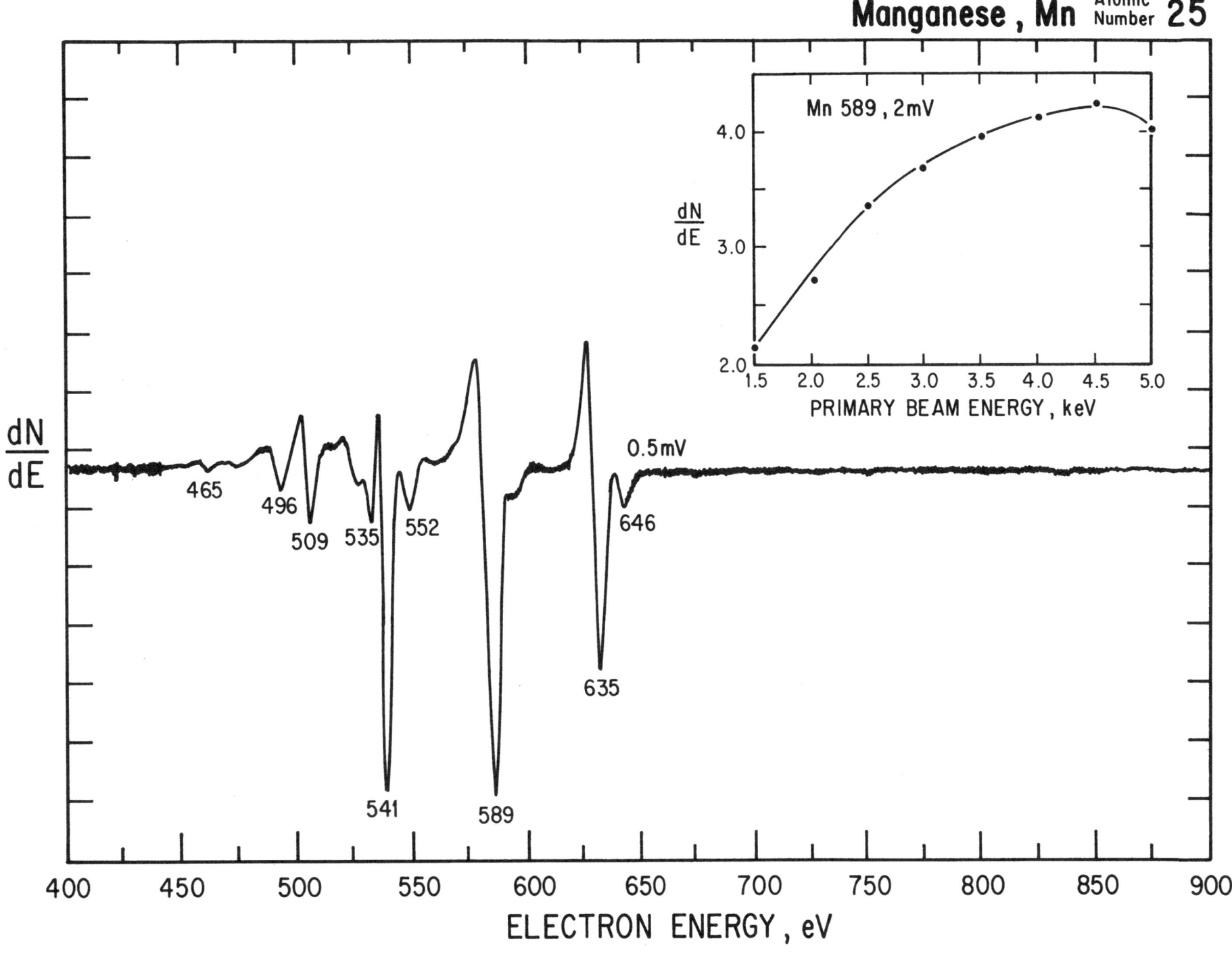

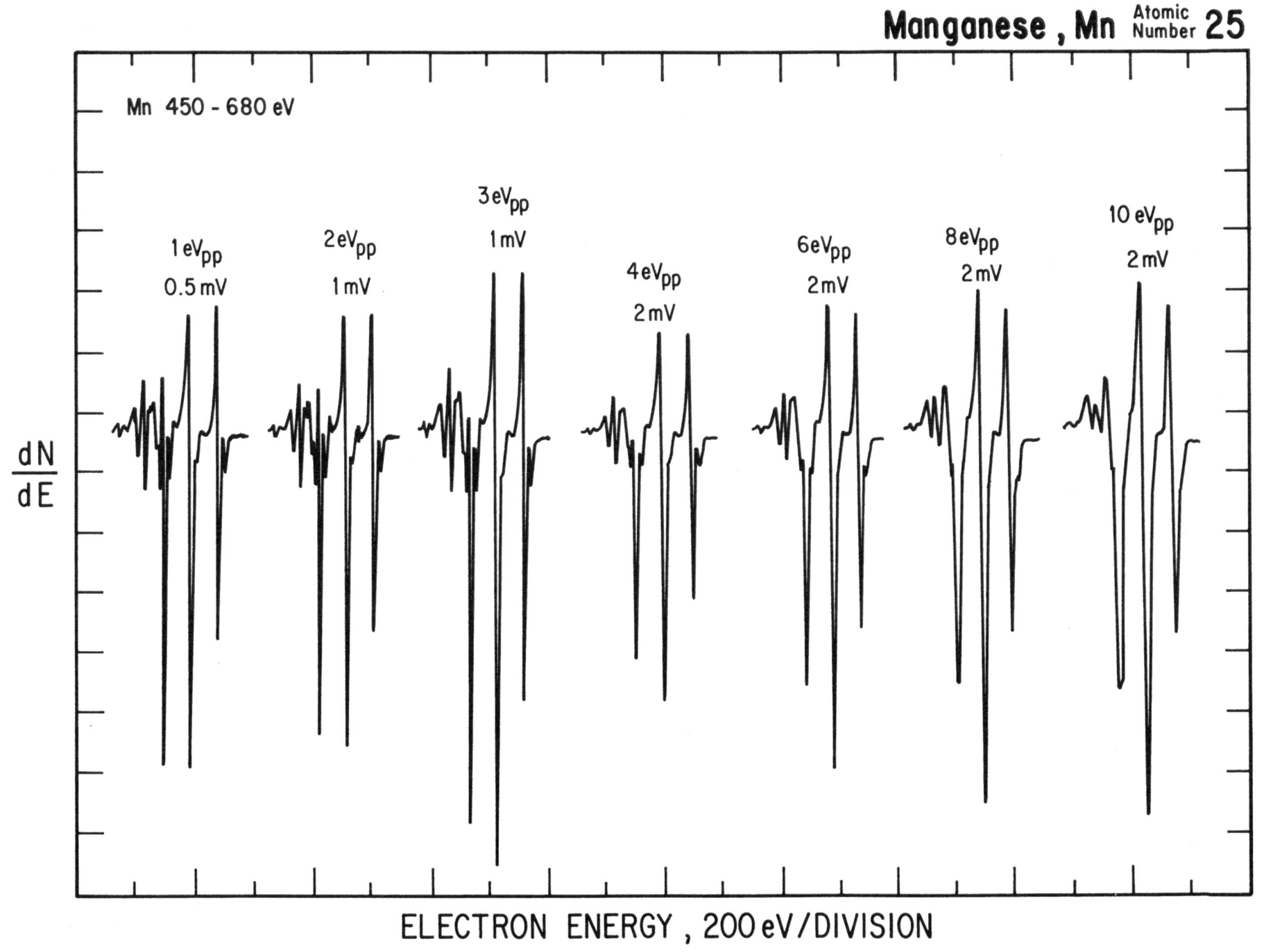

Manganese , Mn Atomic Number **25**

Mn 450 - 680 eV

1 eV$_{pp}$ 0.5 mV

2 eV$_{pp}$ 1 mV

3 eV$_{pp}$ 1 mV

4 eV$_{pp}$ 2 mV

6 eV$_{pp}$ 2 mV

8 eV$_{pp}$ 2 mV

10 eV$_{pp}$ 2 mV

$\dfrac{dN}{dE}$

ELECTRON ENERGY , 200 eV/DIVISION

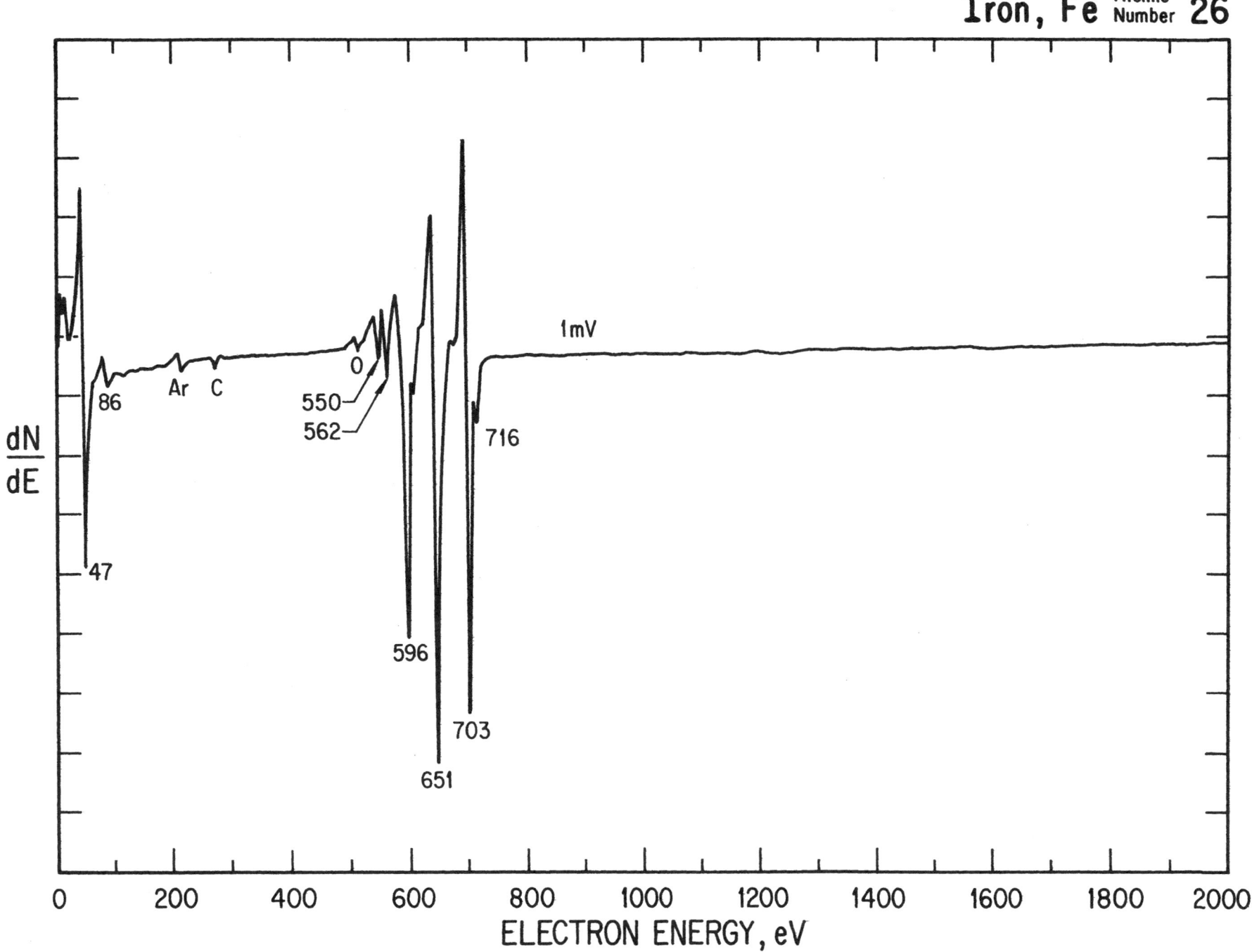

$\dfrac{dN}{dE}$

ELECTRON ENERGY, eV

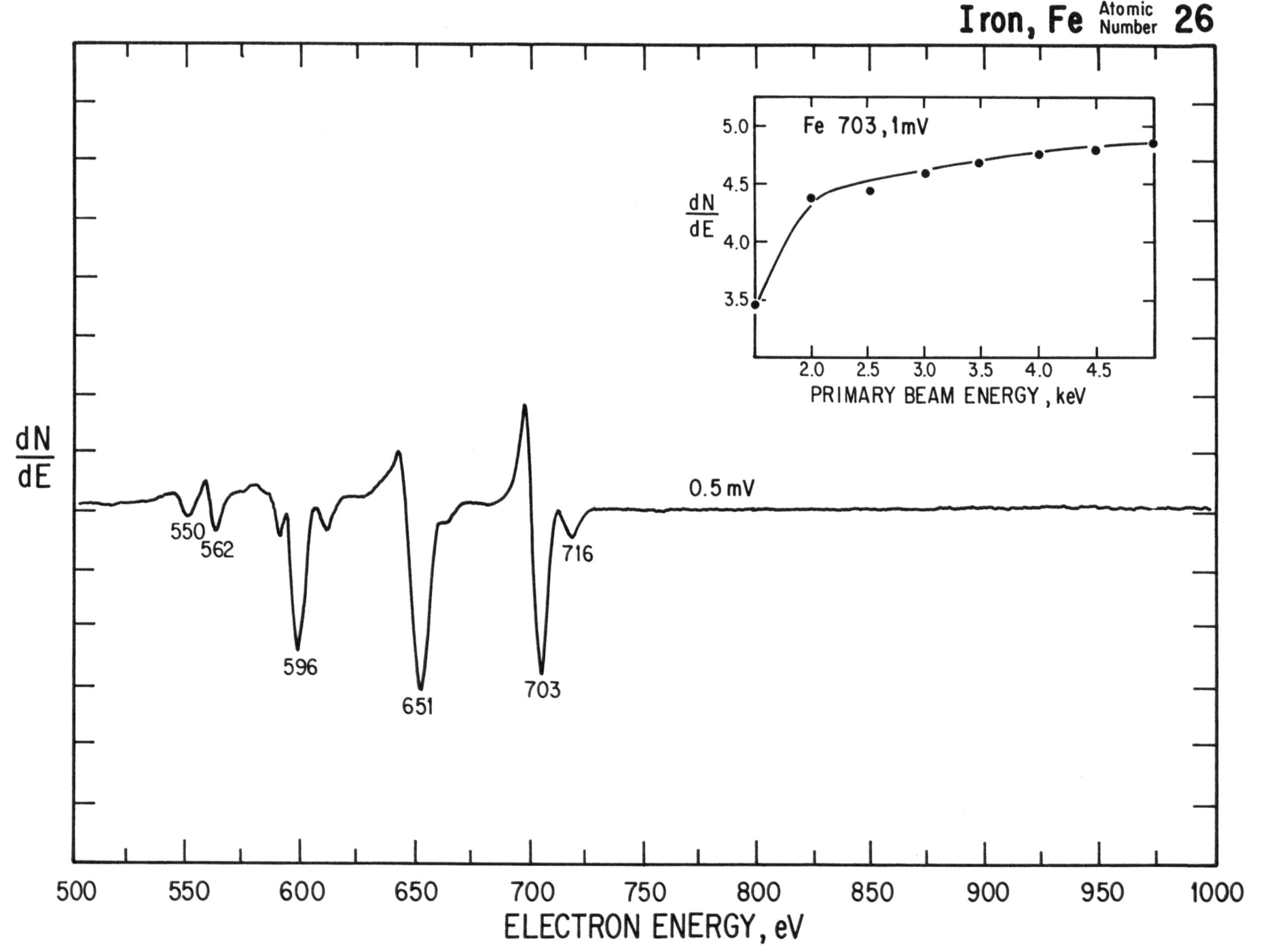

Iron, Fe Atomic Number 26

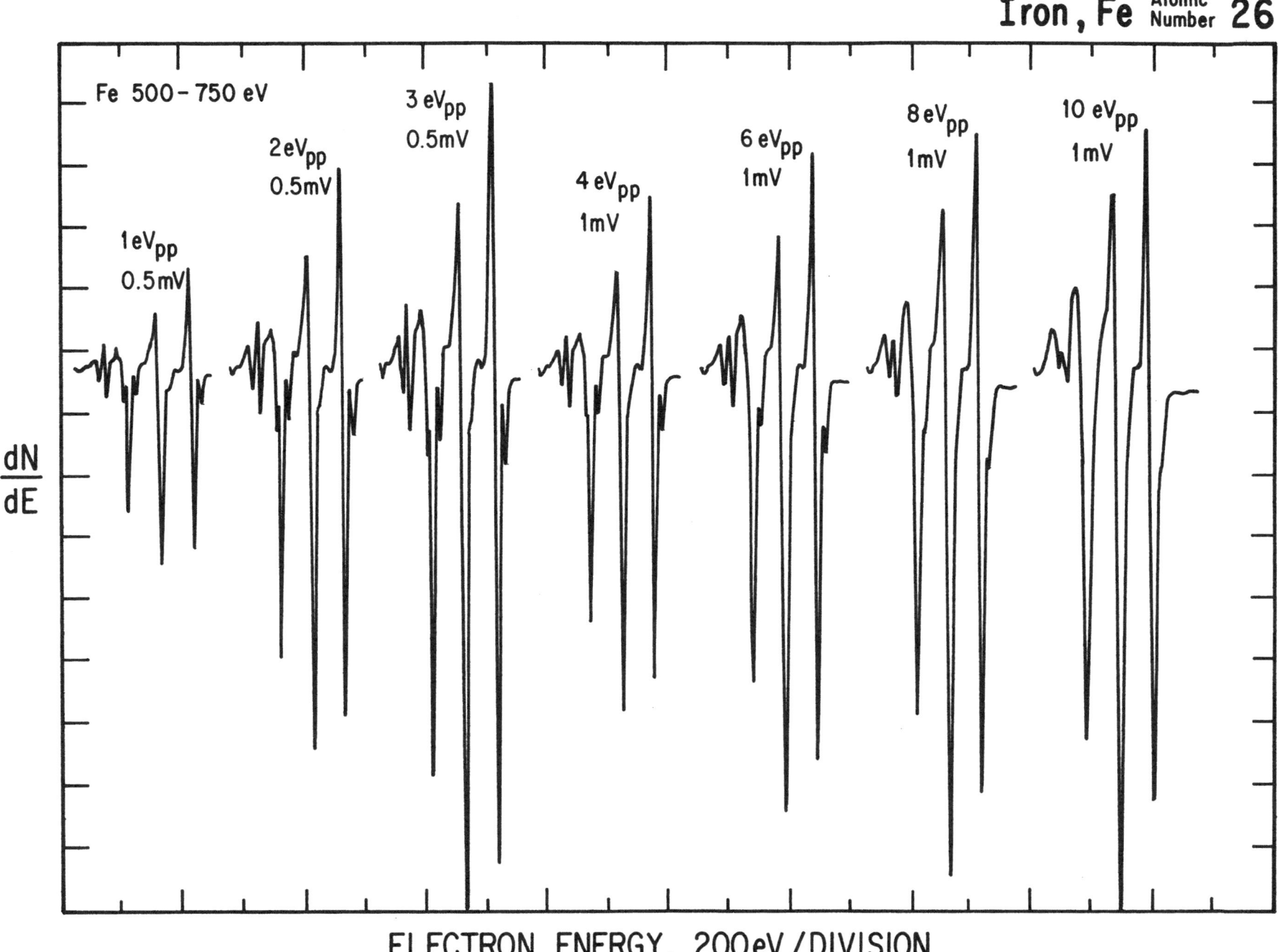

Iron, Fe Atomic Number 26

Fe 500-750 eV

1eV$_{pp}$
0.5mV

2eV$_{pp}$
0.5mV

3 eV$_{pp}$
0.5mV

4 eV$_{pp}$
1mV

6 eV$_{pp}$
1mV

8 eV$_{pp}$
1mV

10 eV$_{pp}$
1mV

$\dfrac{dN}{dE}$

ELECTRON ENERGY, 200eV/DIVISION

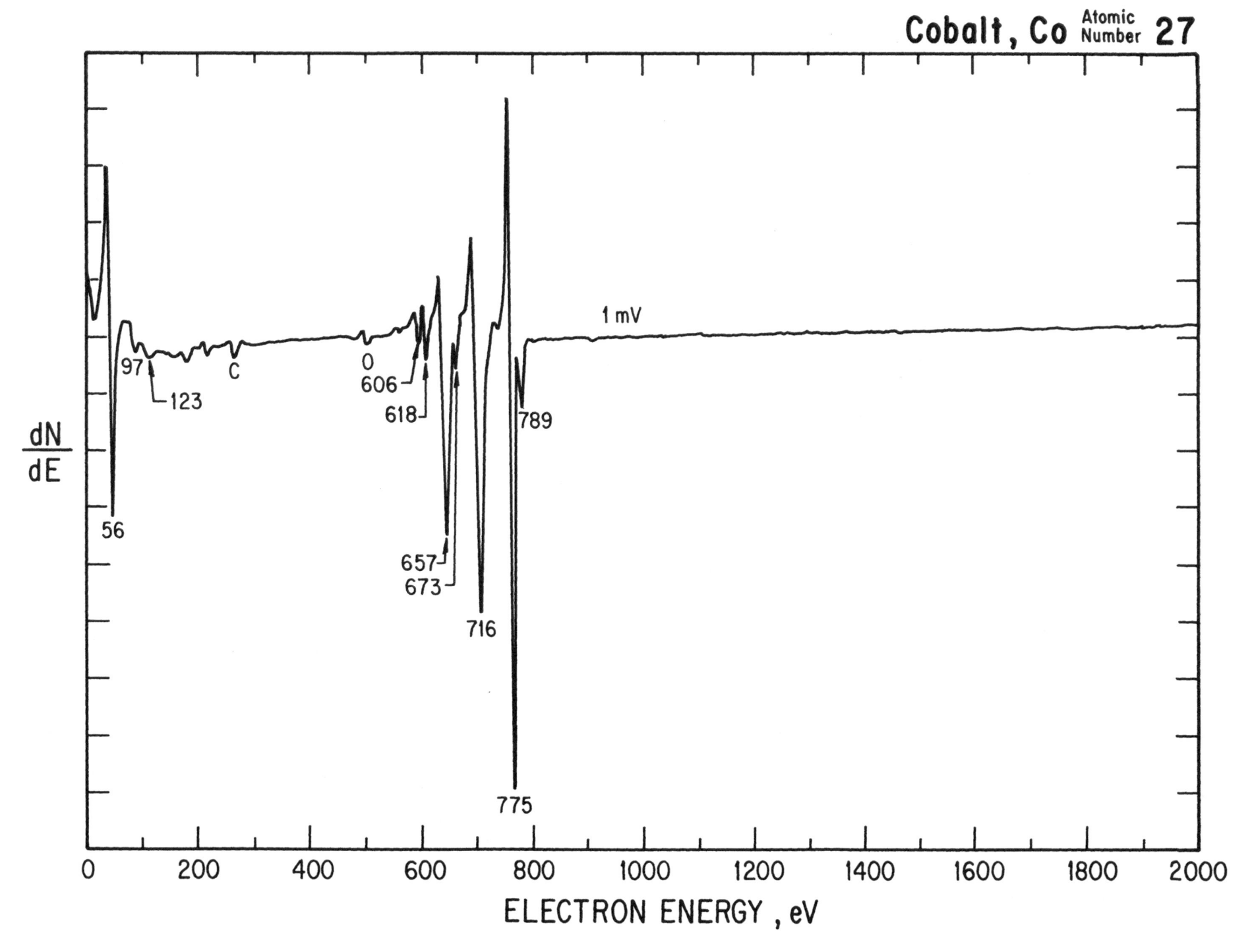

Cobalt, Co Atomic Number 27

dN/dE

ELECTRON ENERGY, eV

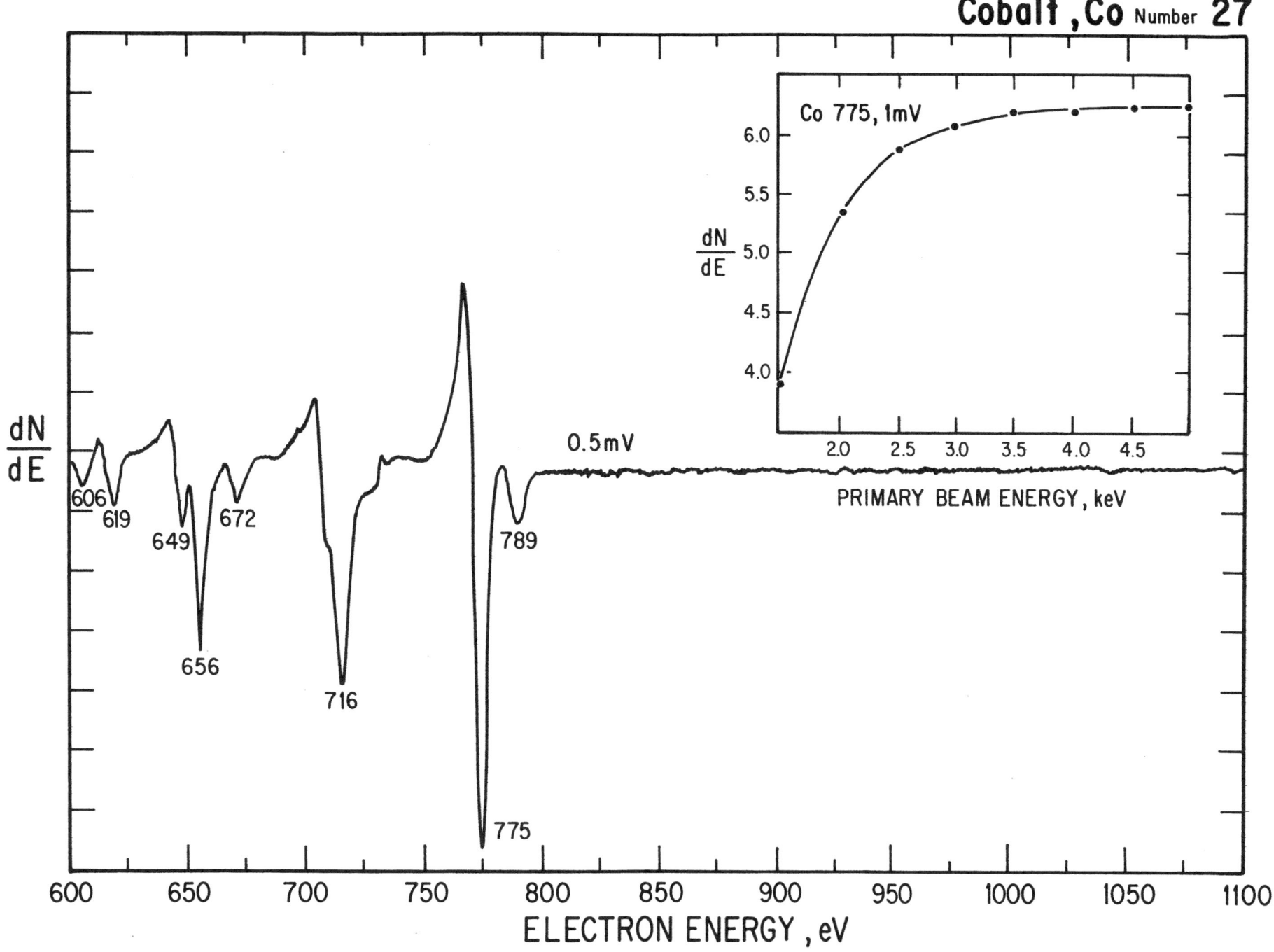

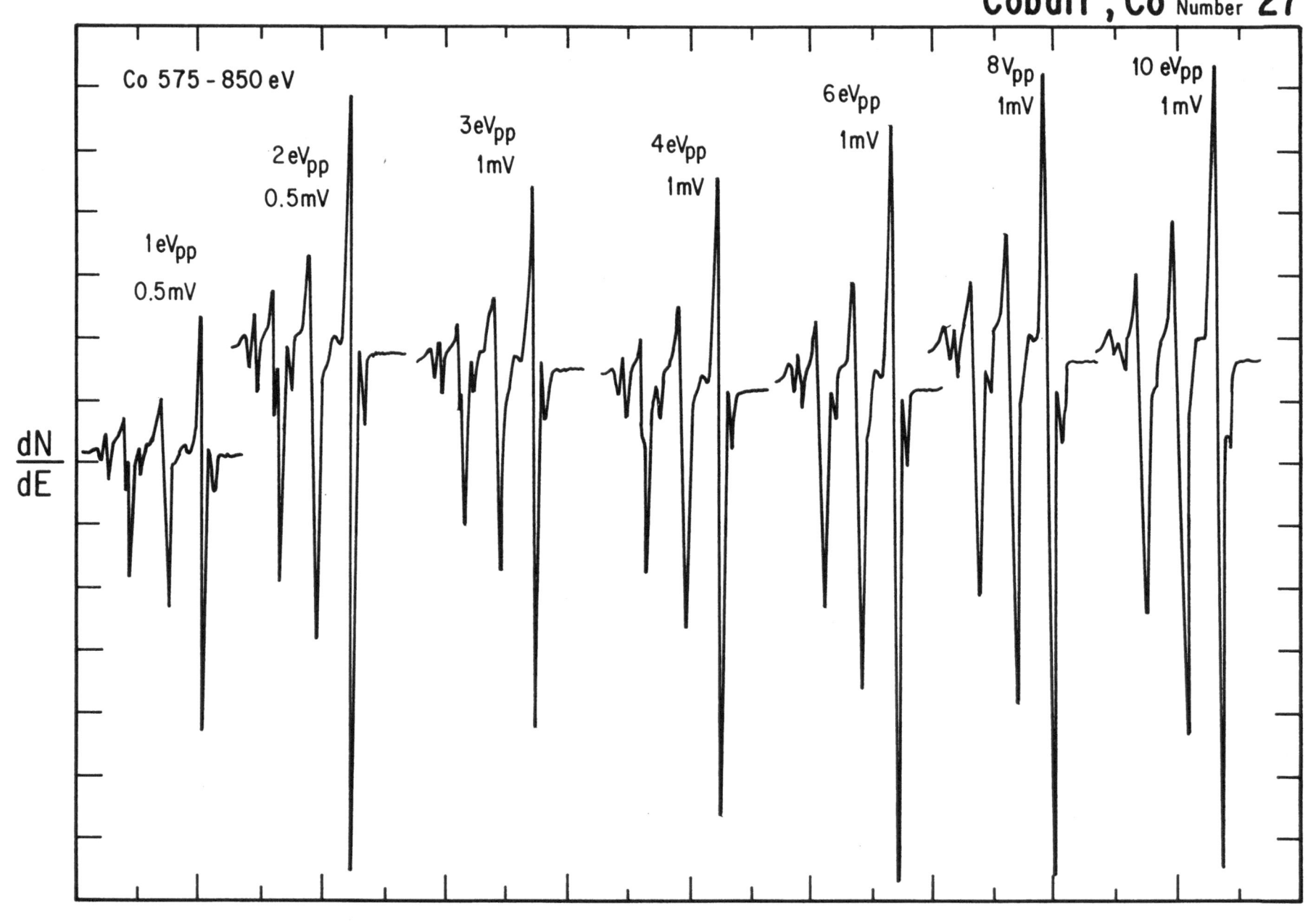

Co 575 - 850 eV

1eV_pp 0.5mV

2eV_pp 0.5mV

3eV_pp 1mV

4eV_pp 1mV

6eV_pp 1mV

8V_pp 1mV

10 eV_pp 1mV

$\frac{dN}{dE}$

ELECTRON ENERGY , 200 eV/DIVISION

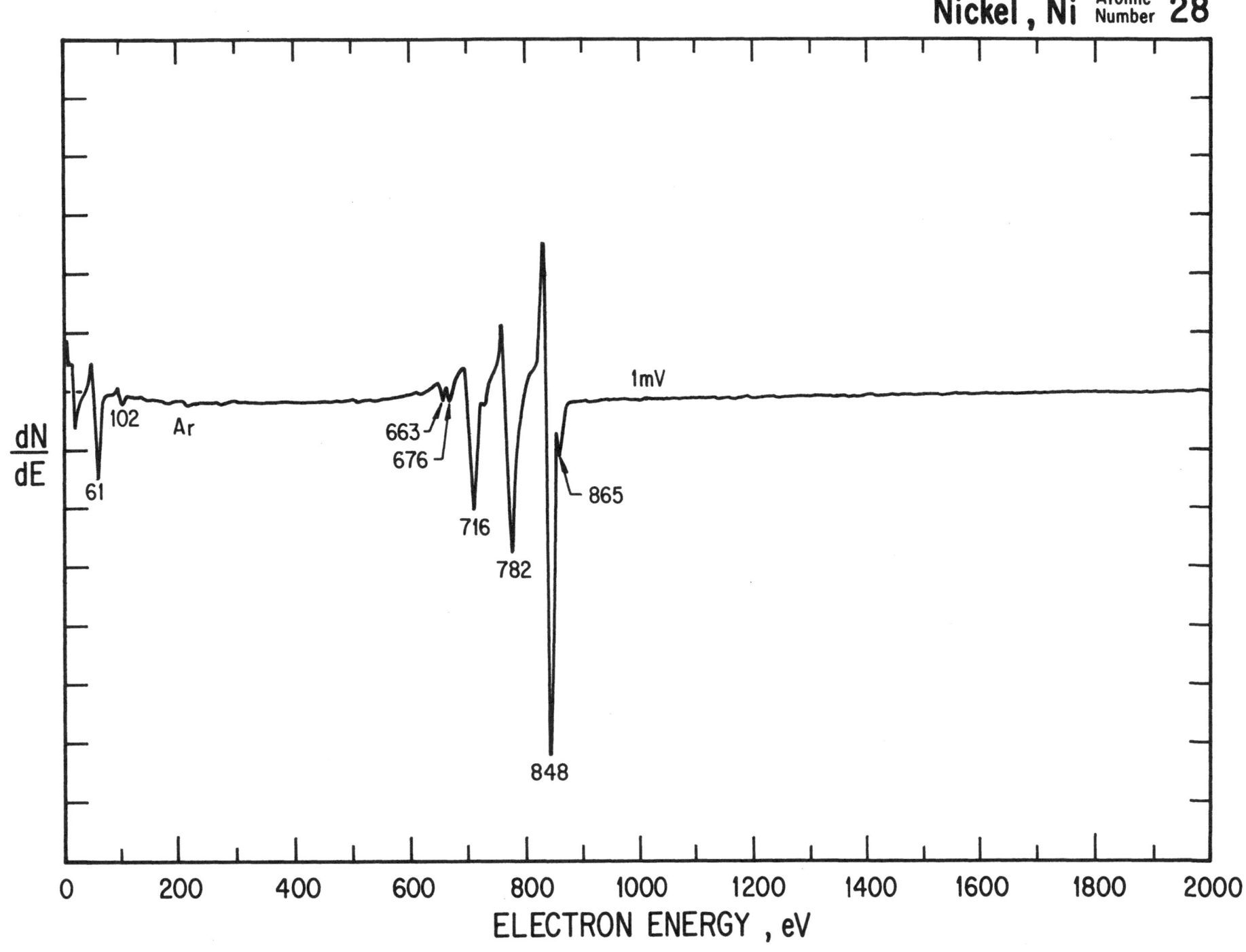

Nickel, Ni   Atomic Number 28

ELECTRON ENERGY, eV

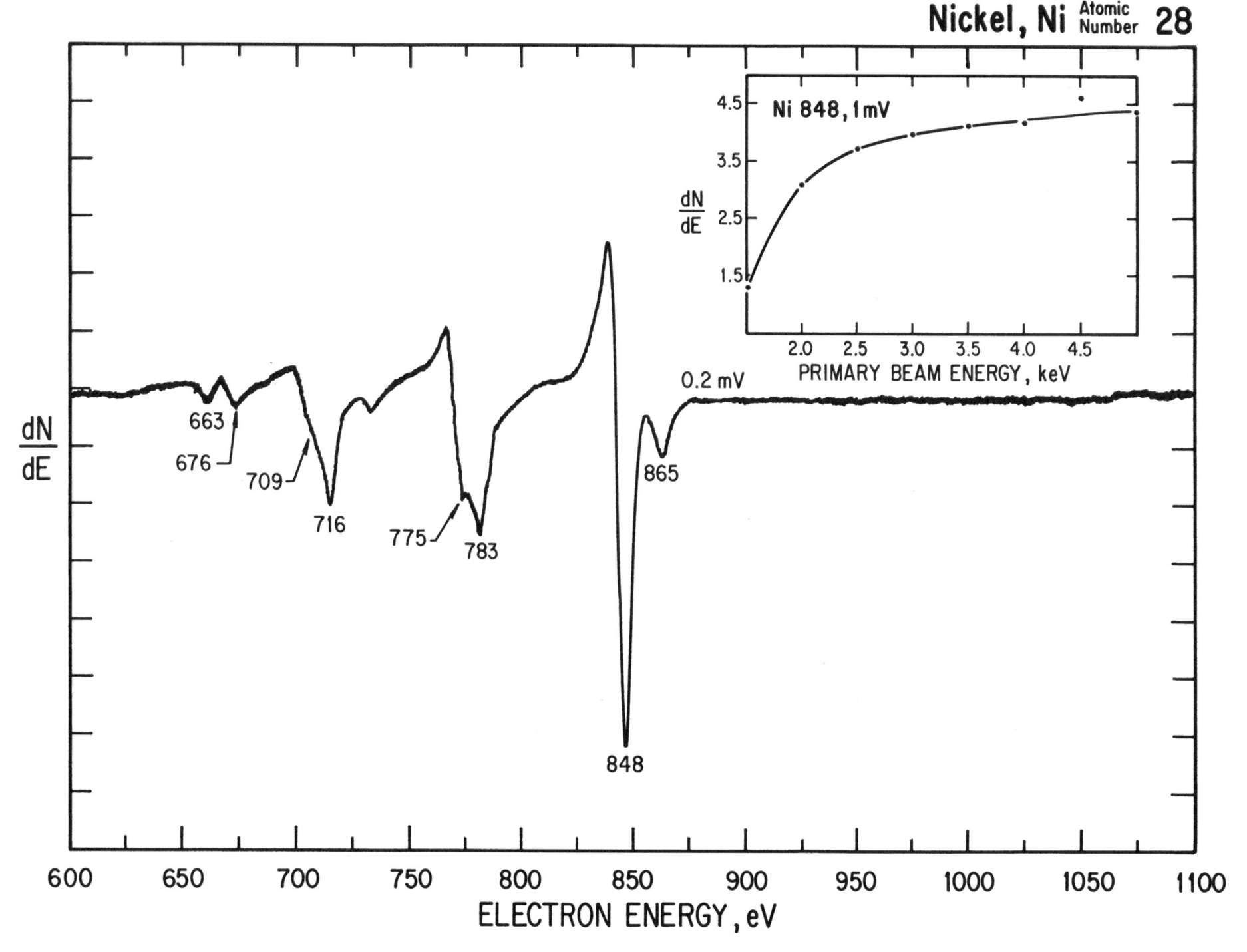

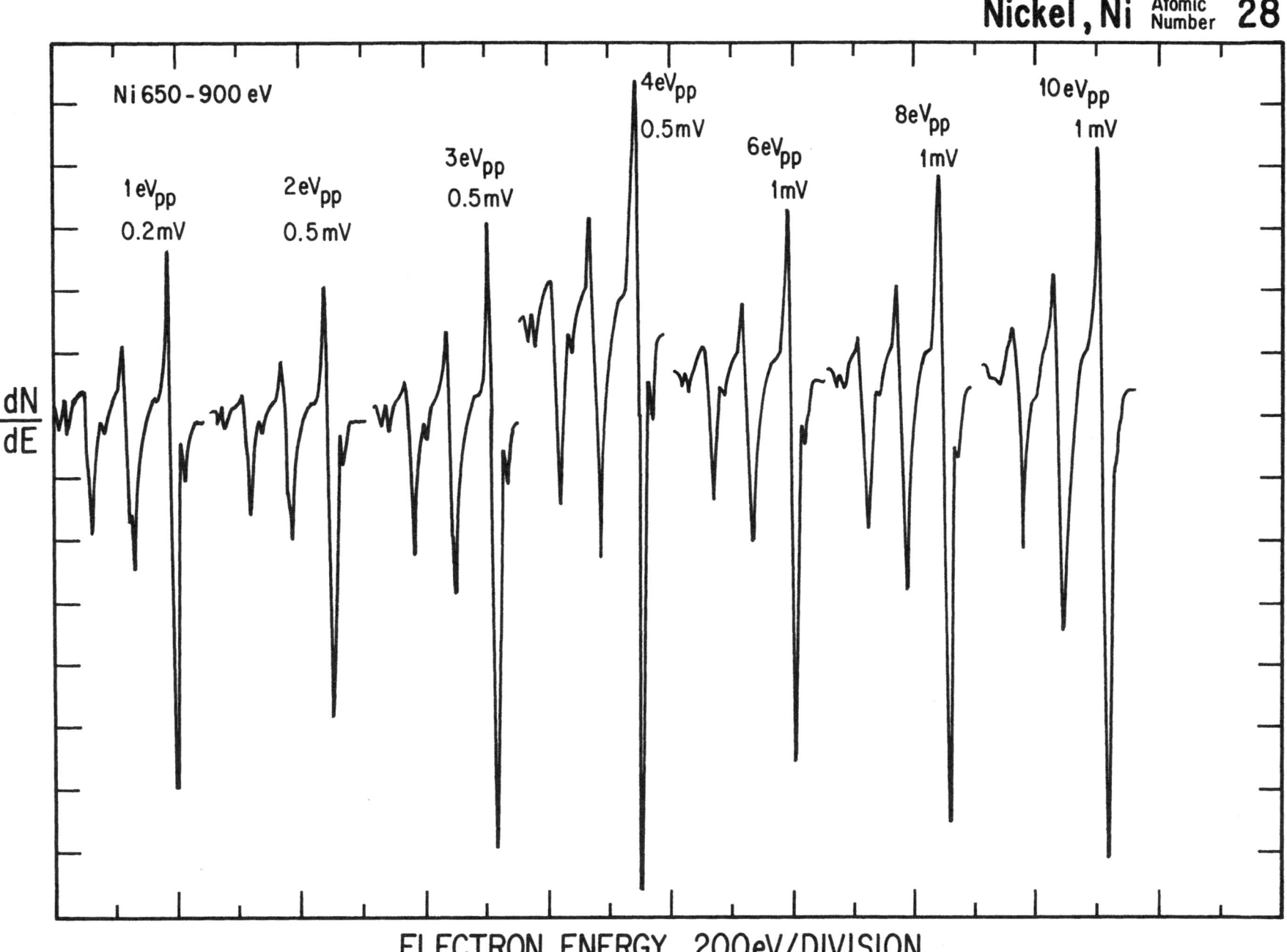

Ni 650 - 900 eV

1 eV$_{pp}$
0.2 mV

2 eV$_{pp}$
0.5 mV

3 eV$_{pp}$
0.5 mV

4 eV$_{pp}$
0.5 mV

6 eV$_{pp}$
1 mV

8 eV$_{pp}$
1 mV

10 eV$_{pp}$
1 mV

$\dfrac{dN}{dE}$

ELECTRON ENERGY, 200 eV/DIVISION

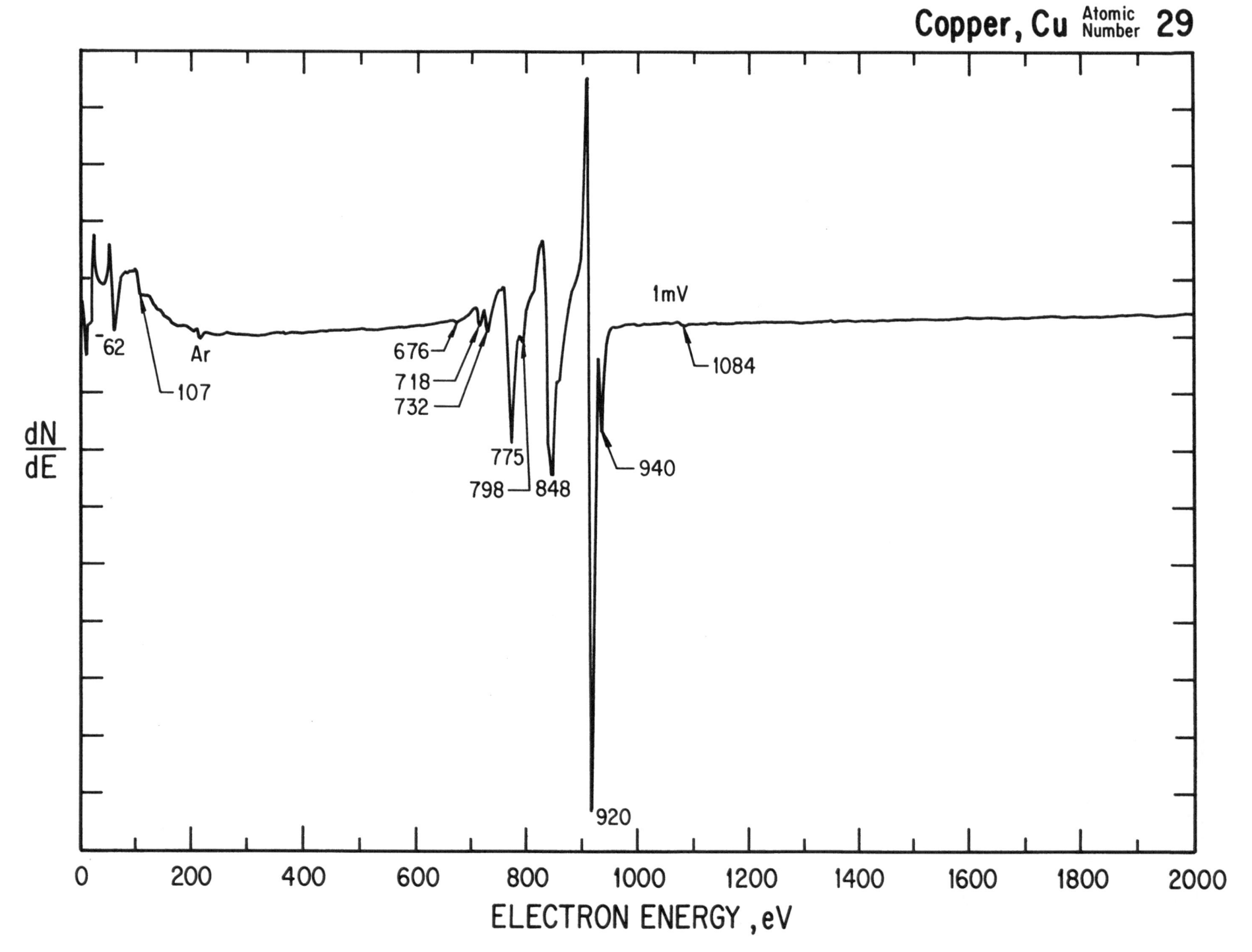

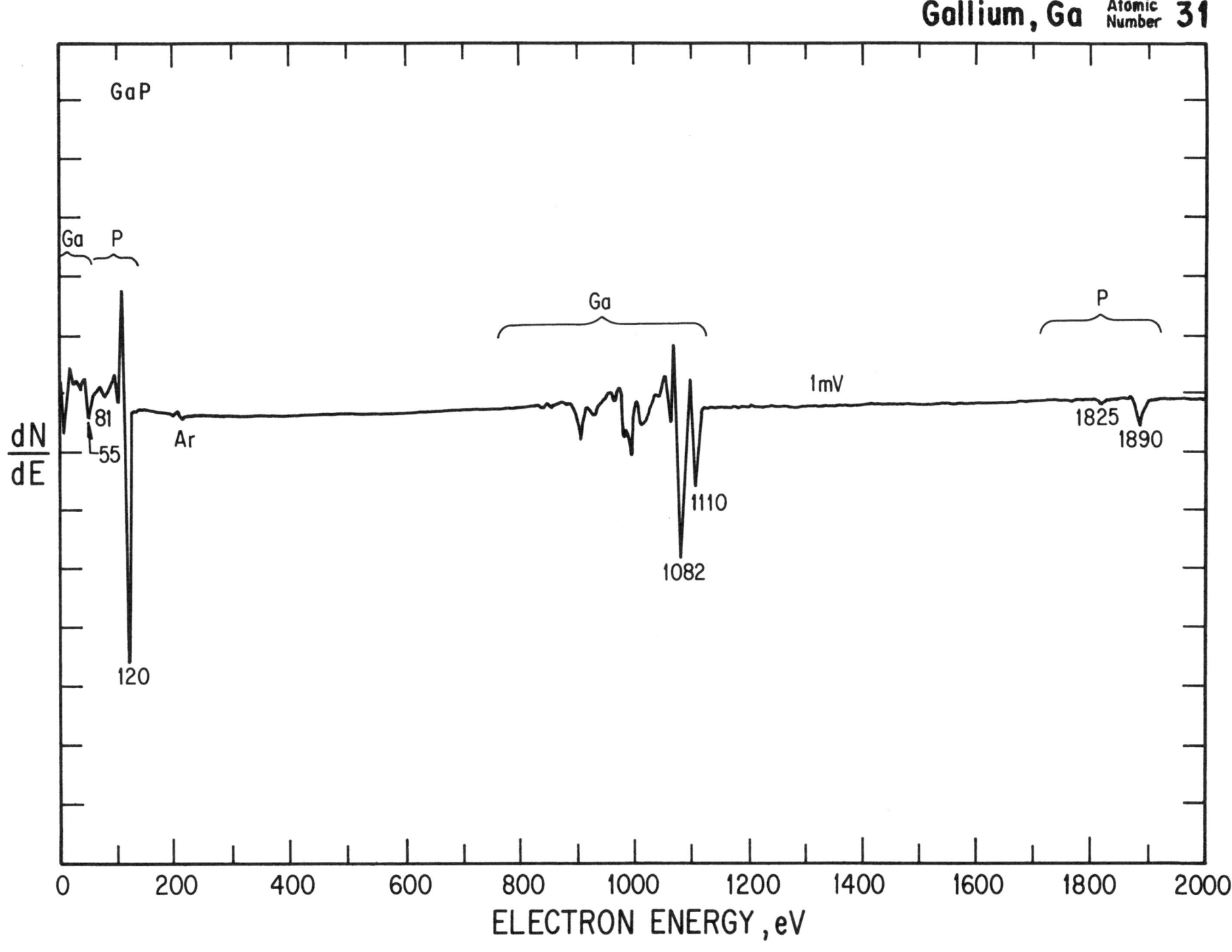

GaP

Ga P

Ga

P

1mV

1825

Ar

81

55

120

1082

1110

1890

$\dfrac{dN}{dE}$

ELECTRON ENERGY, eV

0   200   400   600   800   1000   1200   1400   1600   1800   2000

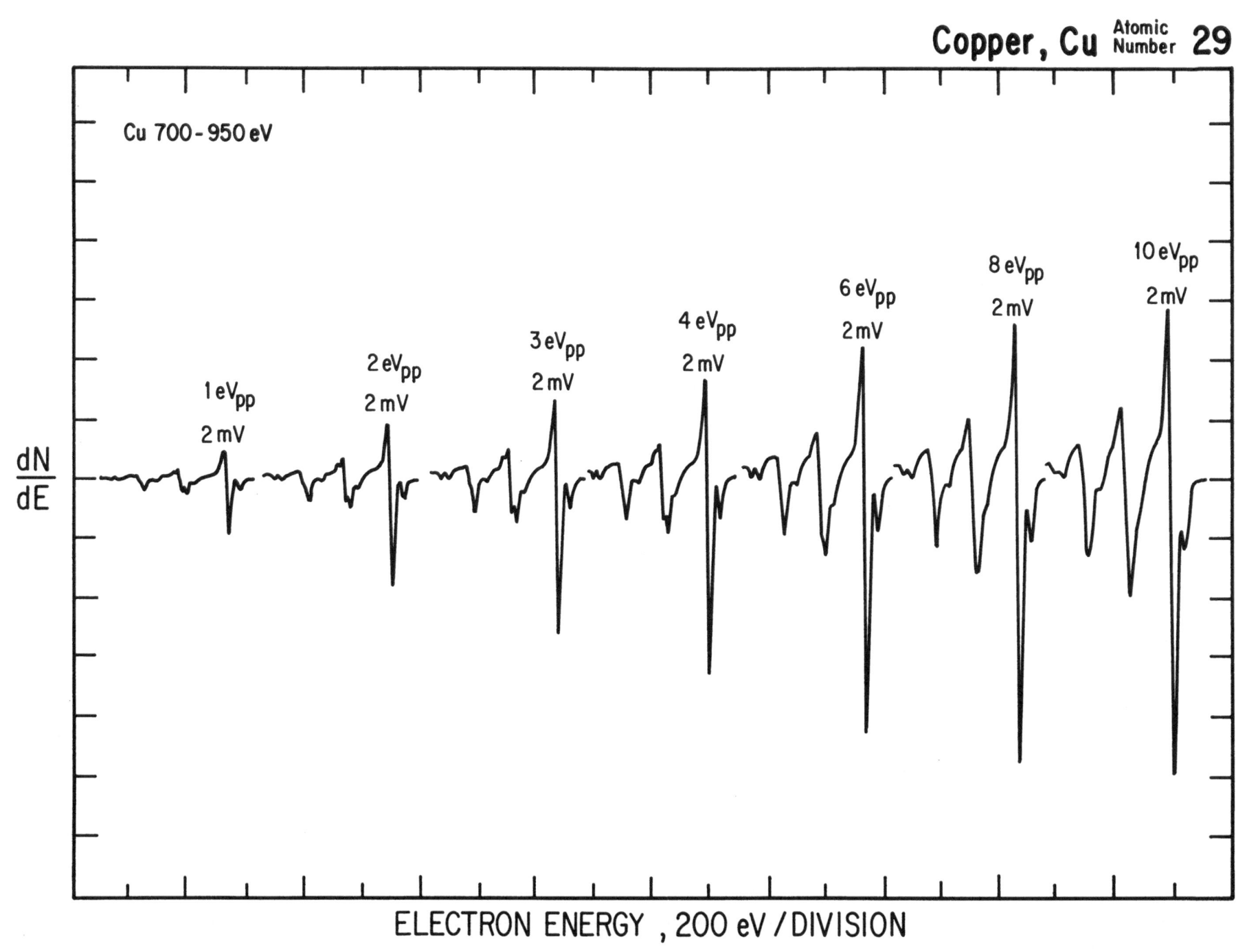

Cu 700-950 eV

1 eV$_{pp}$ 2 mV

2 eV$_{pp}$ 2 mV

3 eV$_{pp}$ 2 mV

4 eV$_{pp}$ 2 mV

6 eV$_{pp}$ 2 mV

8 eV$_{pp}$ 2 mV

10 eV$_{pp}$ 2 mV

$\dfrac{dN}{dE}$

ELECTRON ENERGY , 200 eV / DIVISION

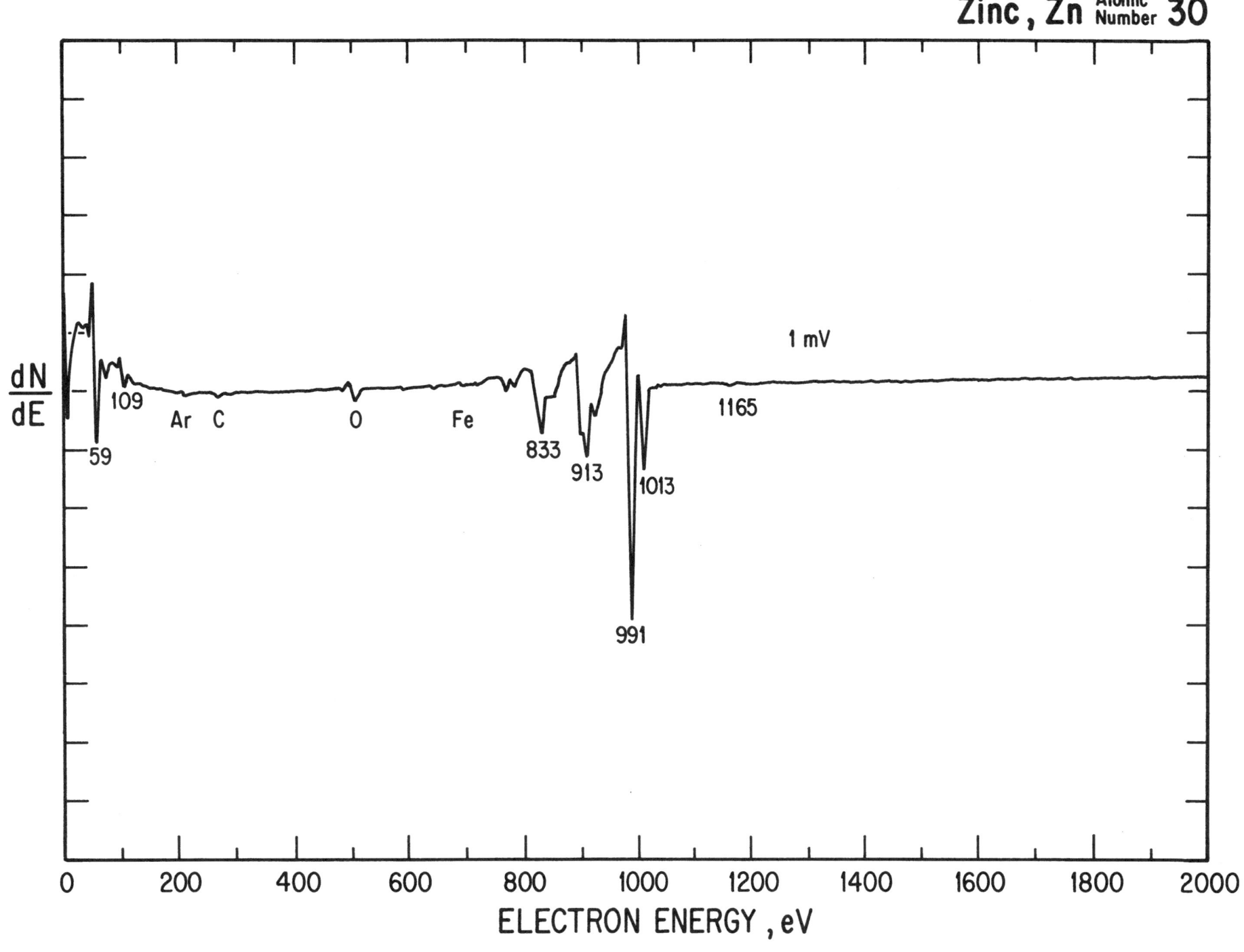

Zinc, Zn Atomic Number 30

$\dfrac{dN}{dE}$

59
109
Ar   C
O
Fe
833
913
991
1013
1165
1 mV

ELECTRON ENERGY, eV

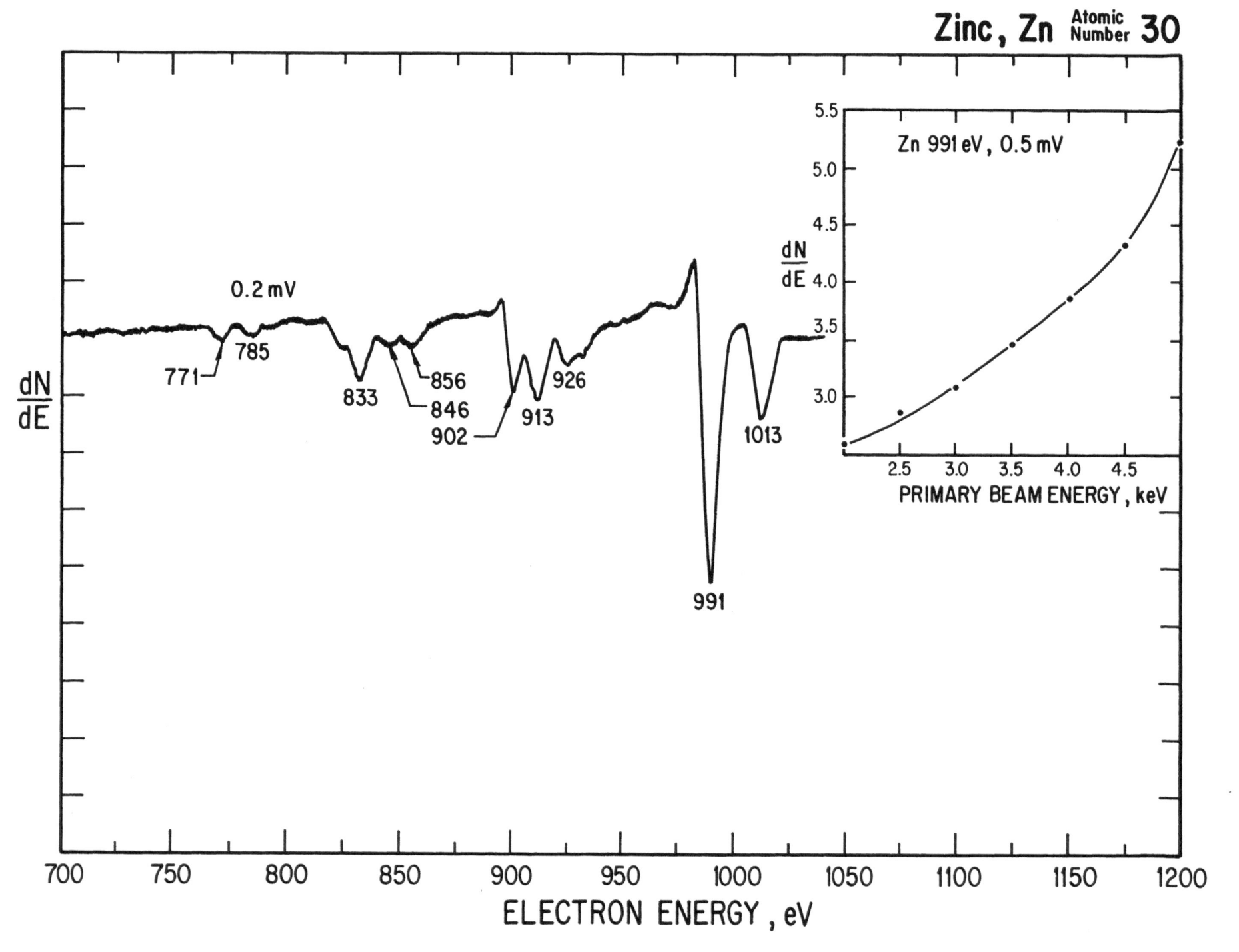

0.2 mV

771

785

833

856

846

902

913

926

991

1013

dN/dE

ELECTRON ENERGY, eV

Zn 991 eV, 0.5 mV

dN/dE

PRIMARY BEAM ENERGY, keV

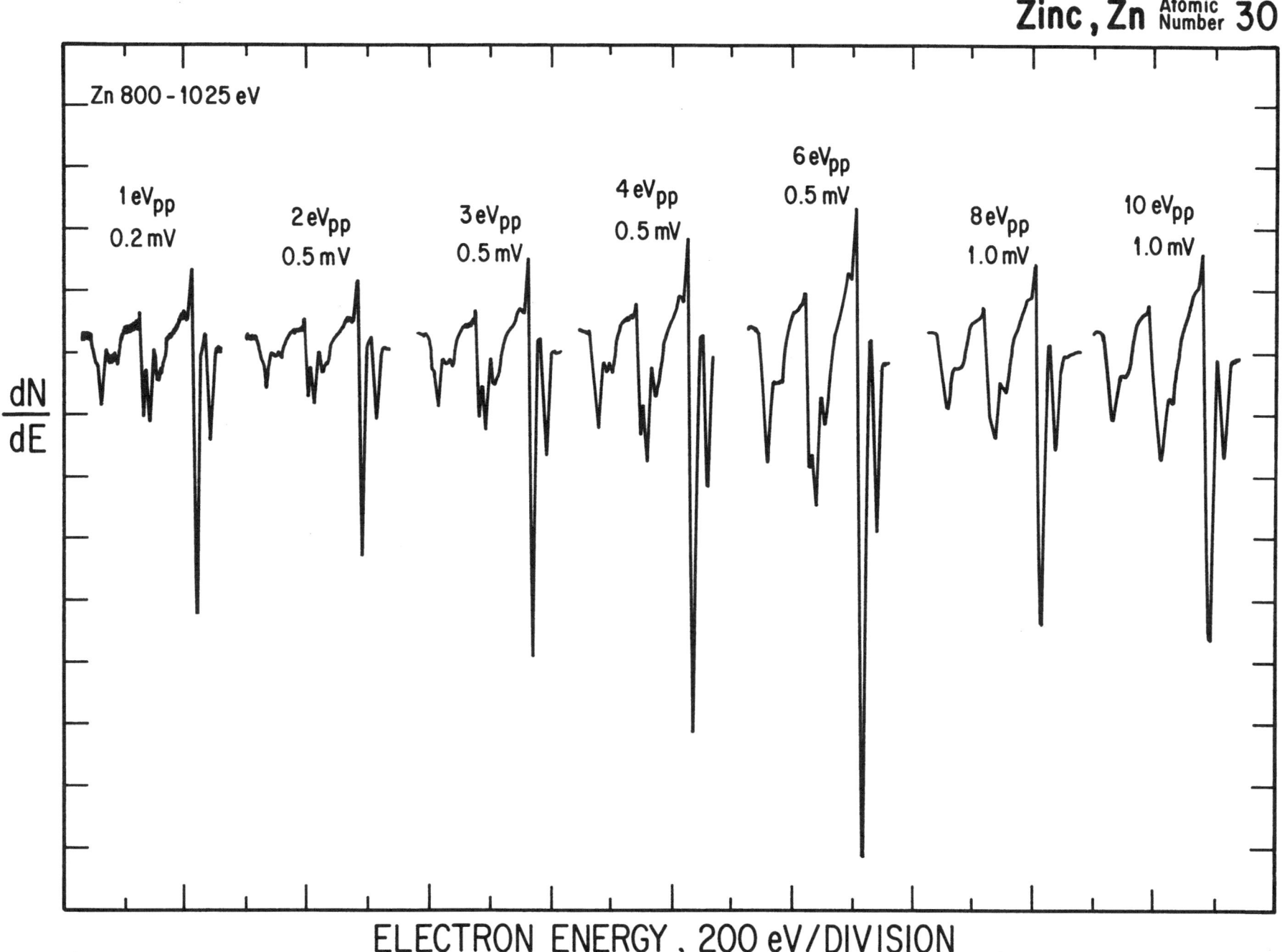

Zn 800 - 1025 eV

1 eV$_{pp}$
0.2 mV

2 eV$_{pp}$
0.5 mV

3 eV$_{pp}$
0.5 mV

4 eV$_{pp}$
0.5 mV

6 eV$_{pp}$
0.5 mV

8 eV$_{pp}$
1.0 mV

10 eV$_{pp}$
1.0 mV

$\dfrac{dN}{dE}$

ELECTRON ENERGY , 200 eV/DIVISION

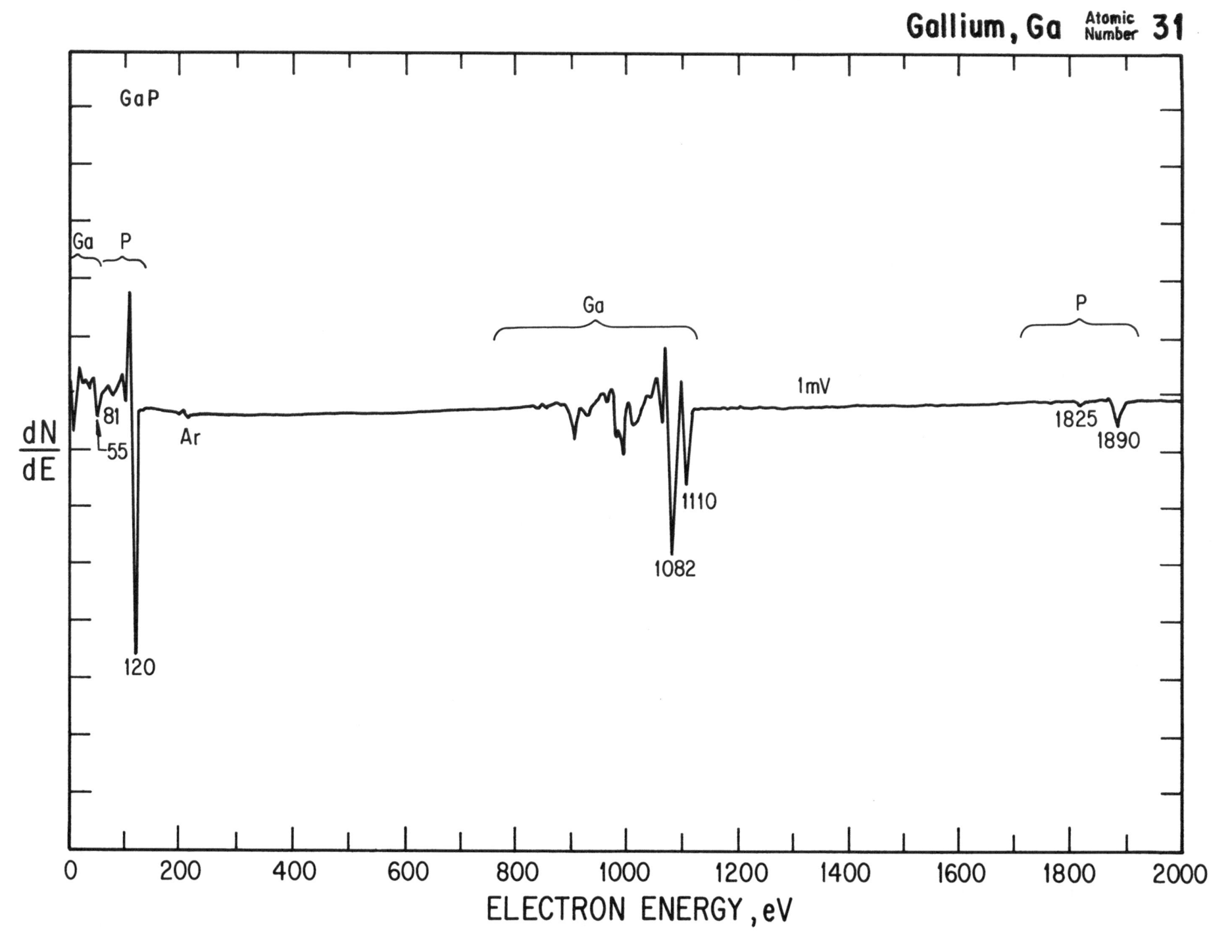

GaP

Ga P

GaP

$\dfrac{dN}{dE}$

81

55

120

Ar

Ga

1082

1110

1mV

P

1825

1890

ELECTRON ENERGY, eV

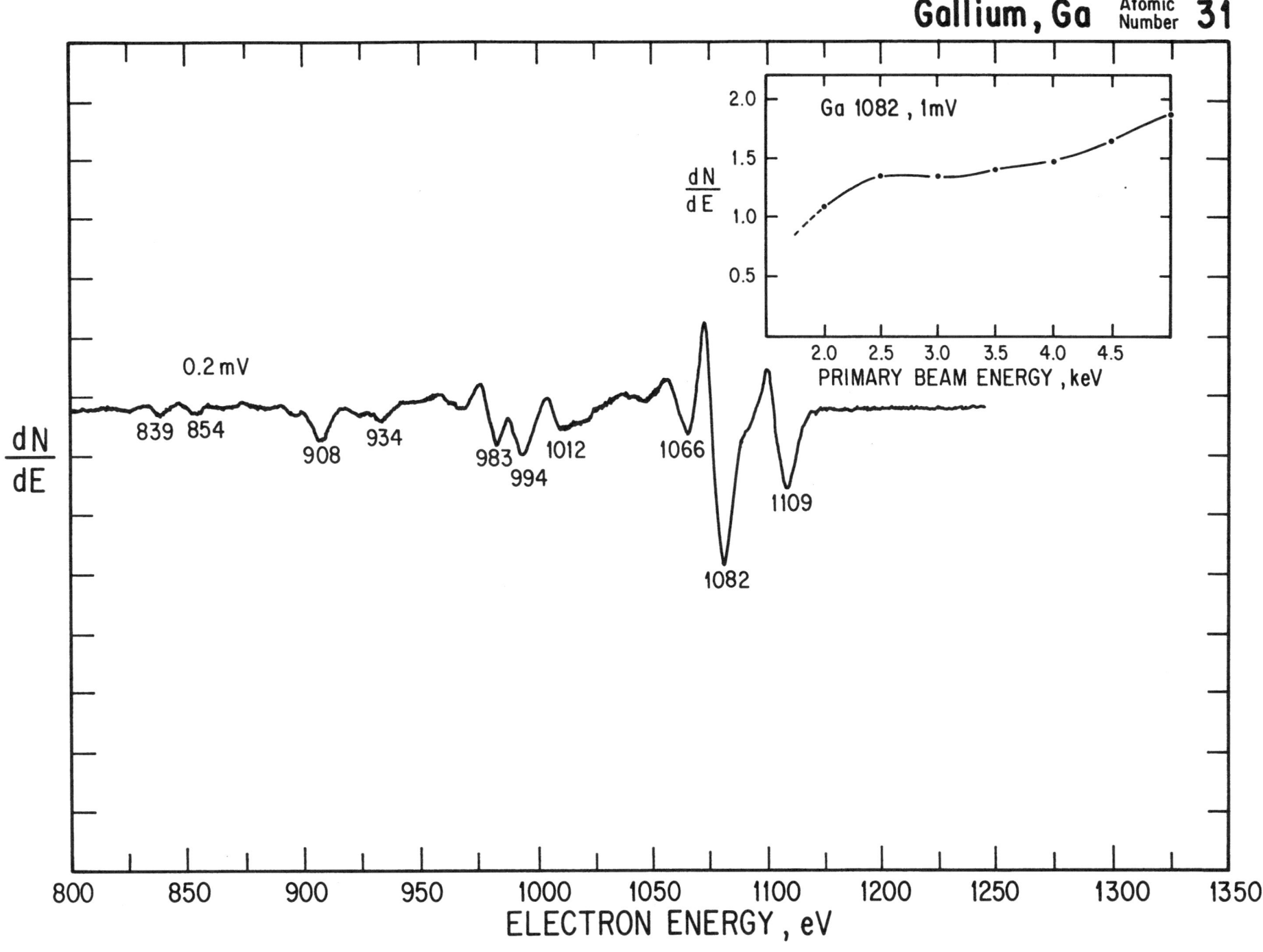

0.2 mV

839 854
908 934
983 1012
994
1066
1082
1109

Ga 1082 , 1mV

$\frac{dN}{dE}$

2.0

1.5

1.0

0.5

2.0 2.5 3.0 3.5 4.0 4.5

PRIMARY BEAM ENERGY , keV

$\frac{dN}{dE}$

ELECTRON ENERGY , eV

800 850 900 950 1000 1050 1100 1200 1250 1300 1350

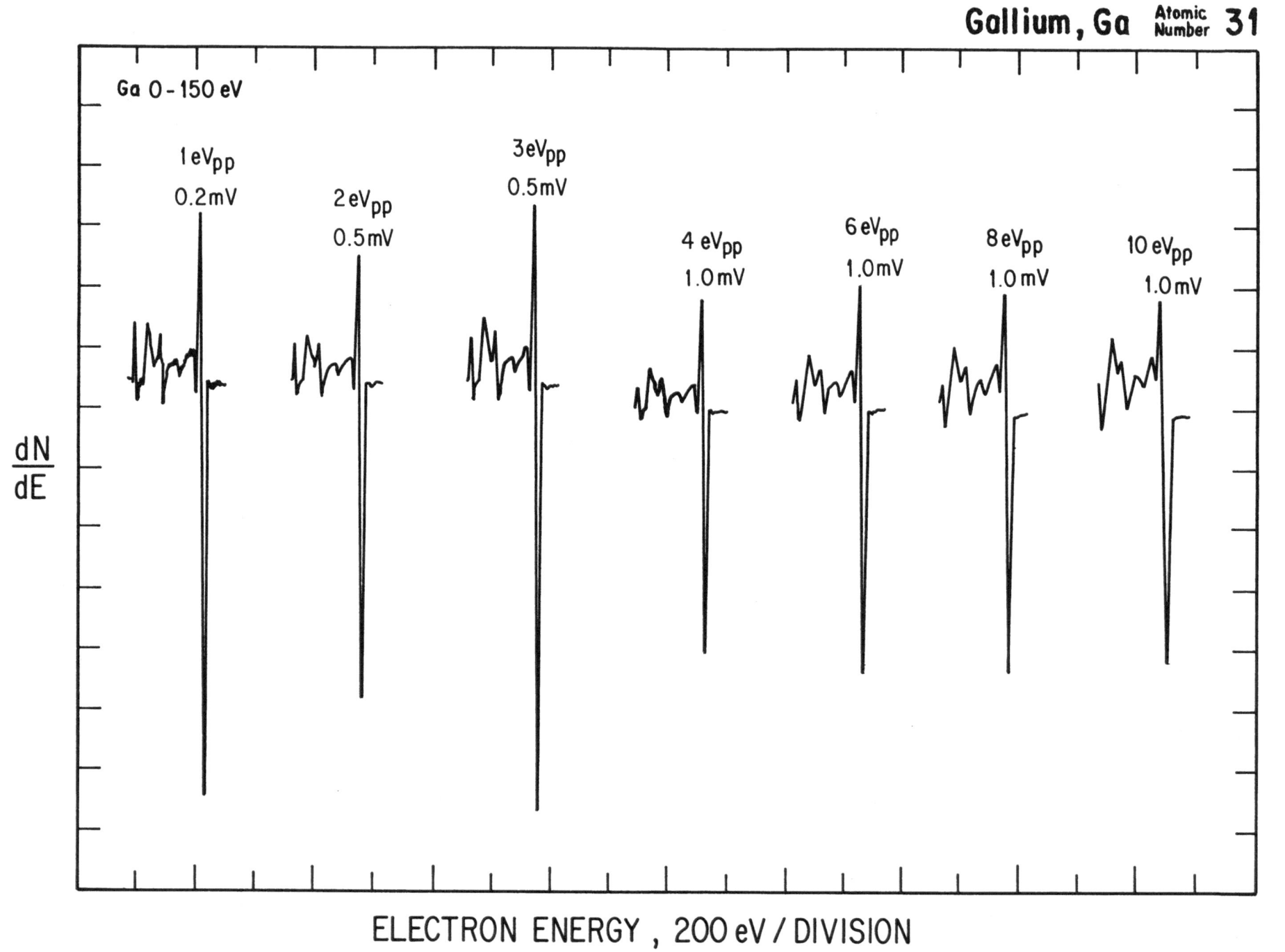

Ga 0 - 150 eV

1eV$_{pp}$
0.2mV

2eV$_{pp}$
0.5mV

3eV$_{pp}$
0.5mV

4 eV$_{pp}$
1.0 mV

6 eV$_{pp}$
1.0mV

8 eV$_{pp}$
1.0 mV

10 eV$_{pp}$
1.0 mV

$\dfrac{dN}{dE}$

ELECTRON ENERGY , 200 eV / DIVISION

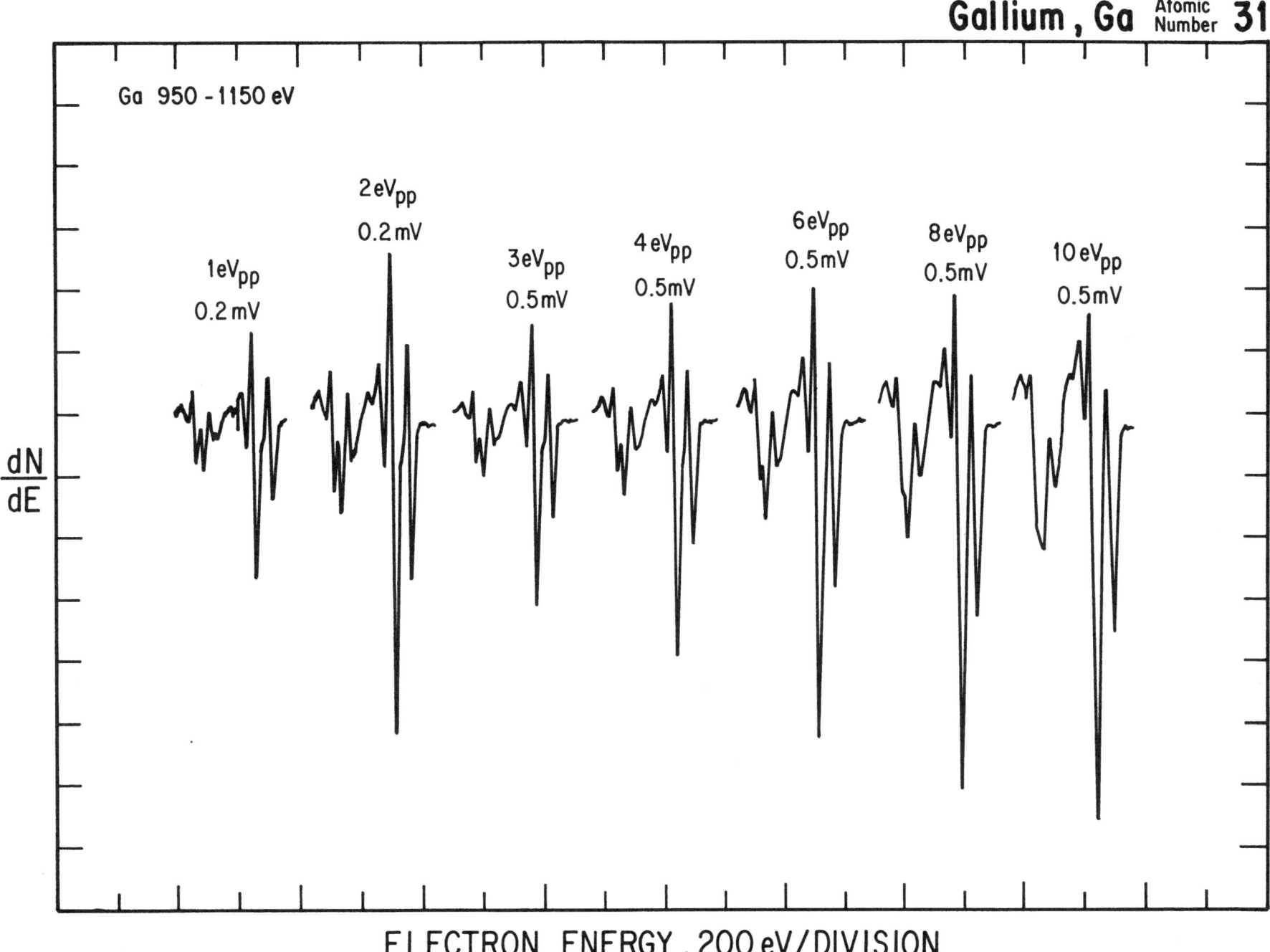

Ga 950 - 1150 eV

1eV_pp 0.2 mV

2eV_pp 0.2 mV

3eV_pp 0.5mV

4eV_pp 0.5mV

6eV_pp 0.5mV

8eV_pp 0.5mV

10eV_pp 0.5mV

$\frac{dN}{dE}$

ELECTRON ENERGY , 200 eV/DIVISION

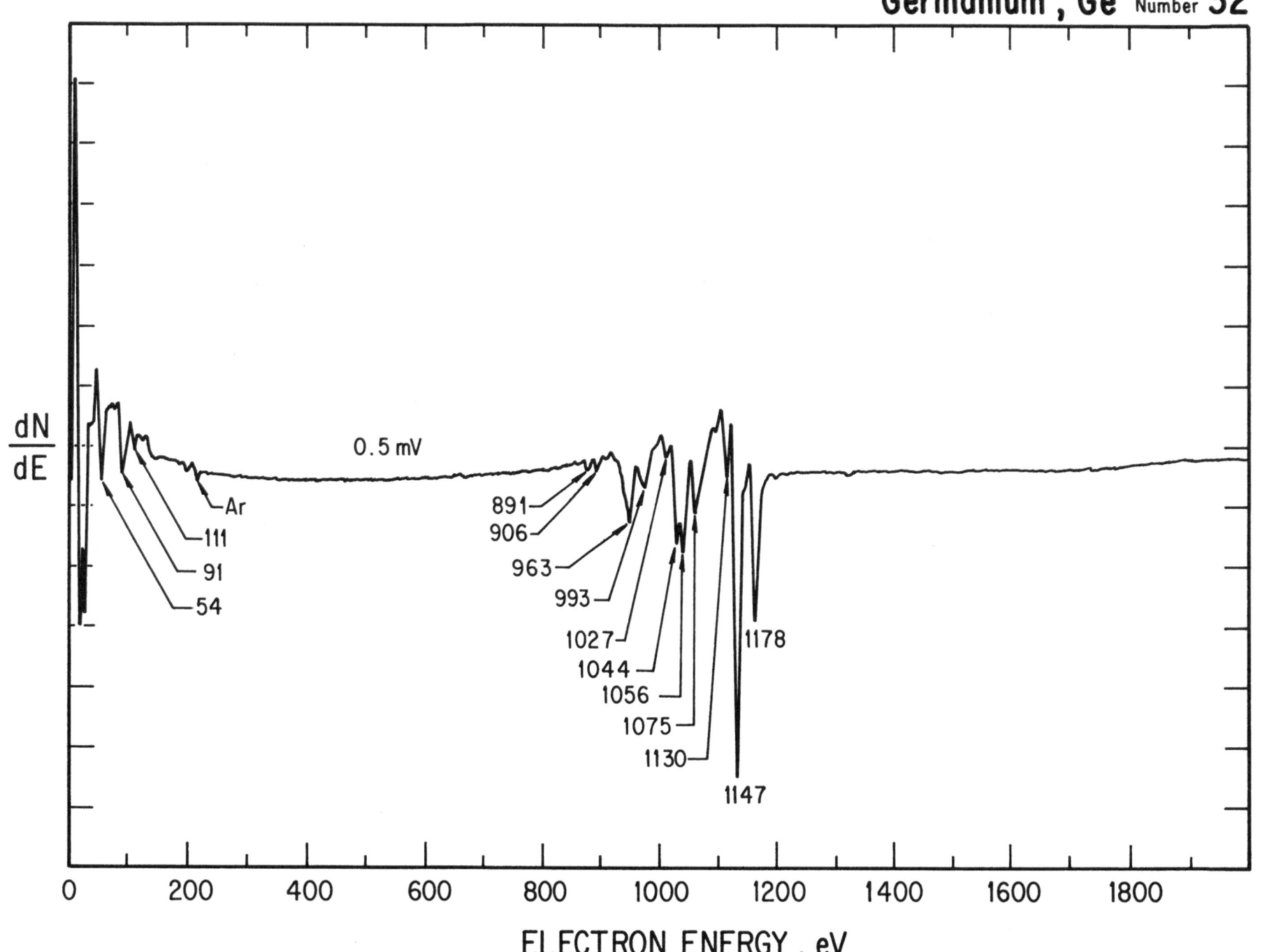

$\dfrac{dN}{dE}$

0.5 mV

Ar
111
91
54

891
906
963
993
1027
1044
1056
1075
1130
1147
1178

ELECTRON ENERGY , eV

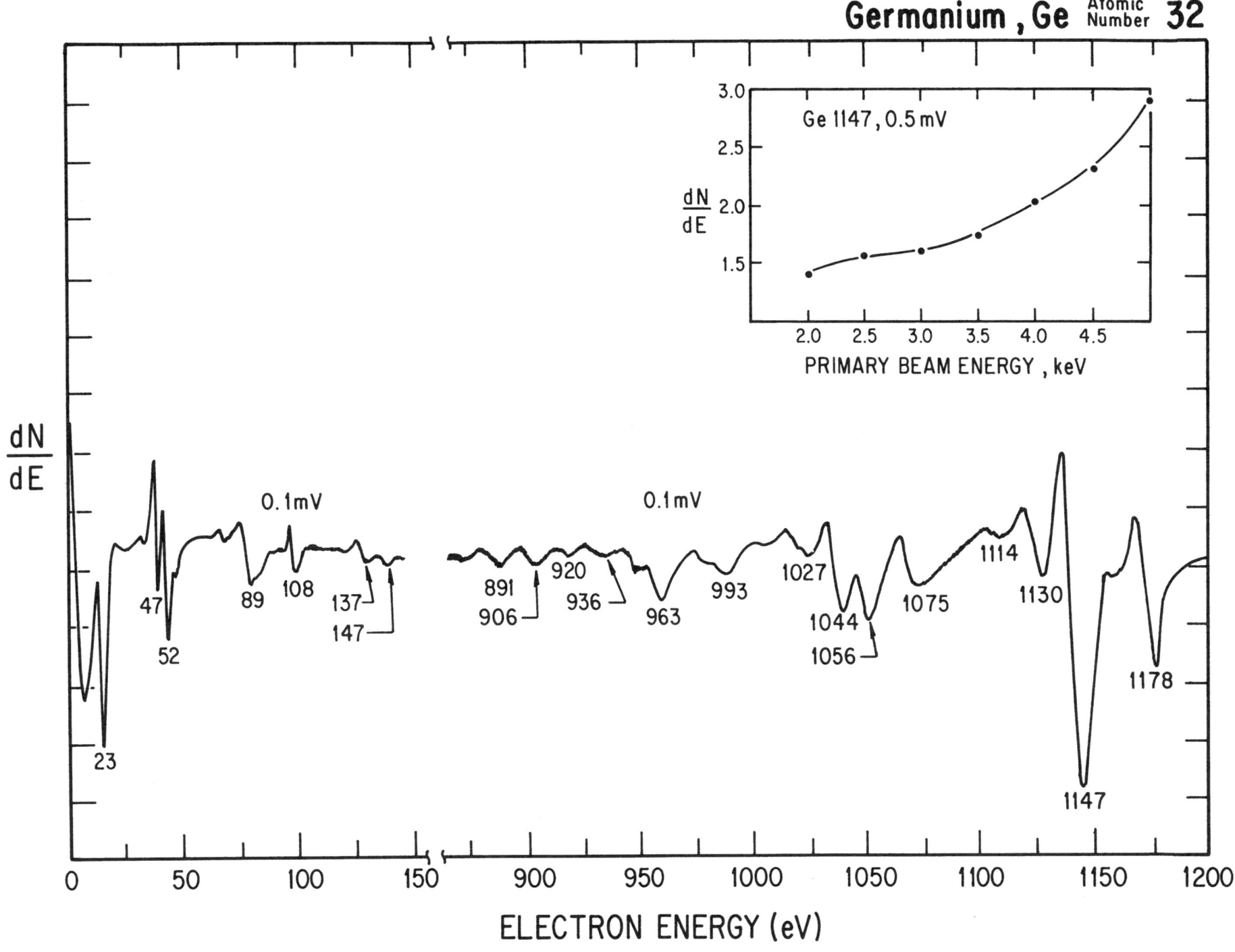

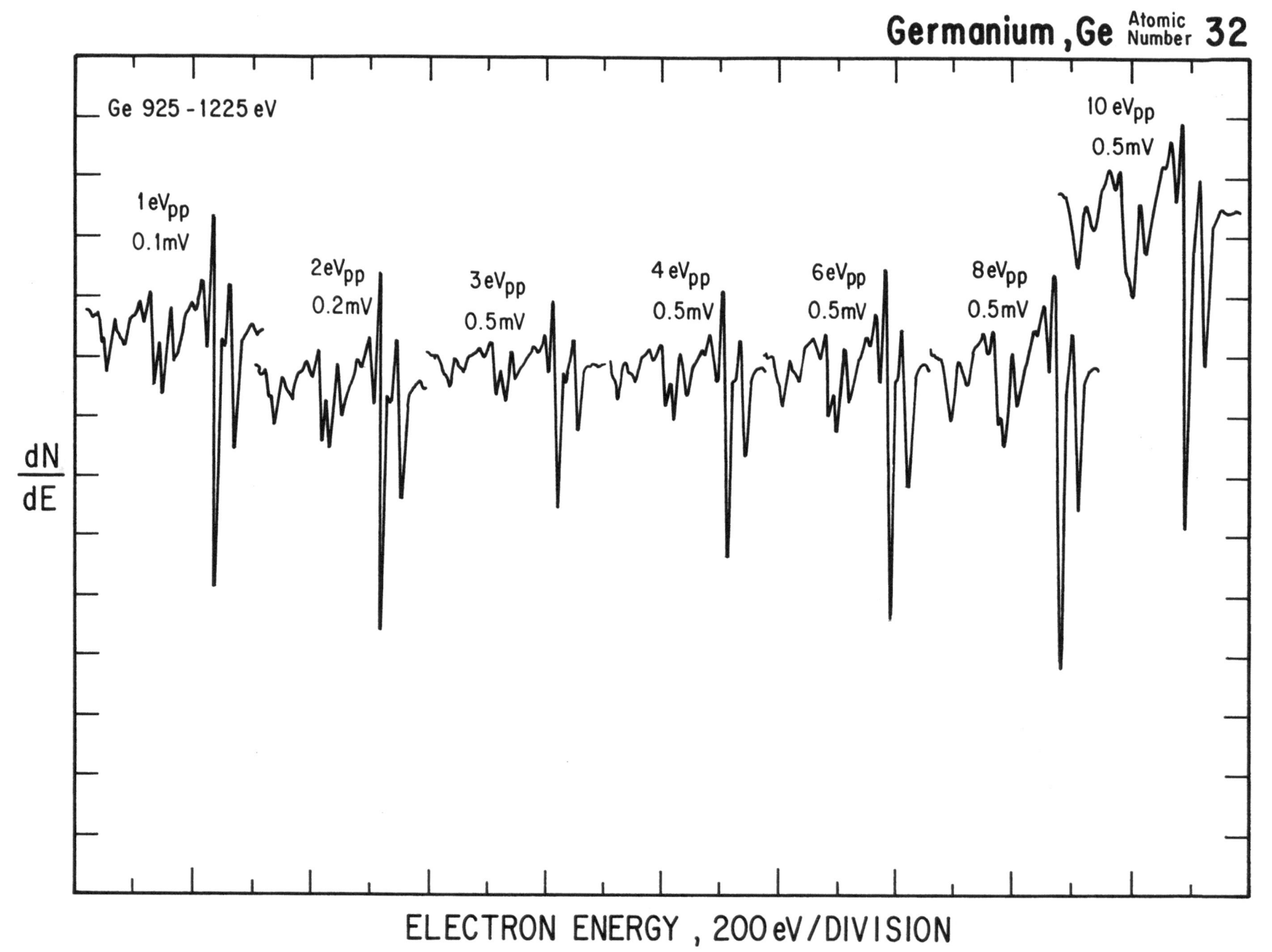

Ge 925 - 1225 eV

1 eV$_{pp}$
0.1mV

2 eV$_{pp}$
0.2mV

3 eV$_{pp}$
0.5mV

4 eV$_{pp}$
0.5mV

6 eV$_{pp}$
0.5mV

8 eV$_{pp}$
0.5mV

10 eV$_{pp}$
0.5mV

$\dfrac{dN}{dE}$

ELECTRON ENERGY , 200 eV/DIVISION

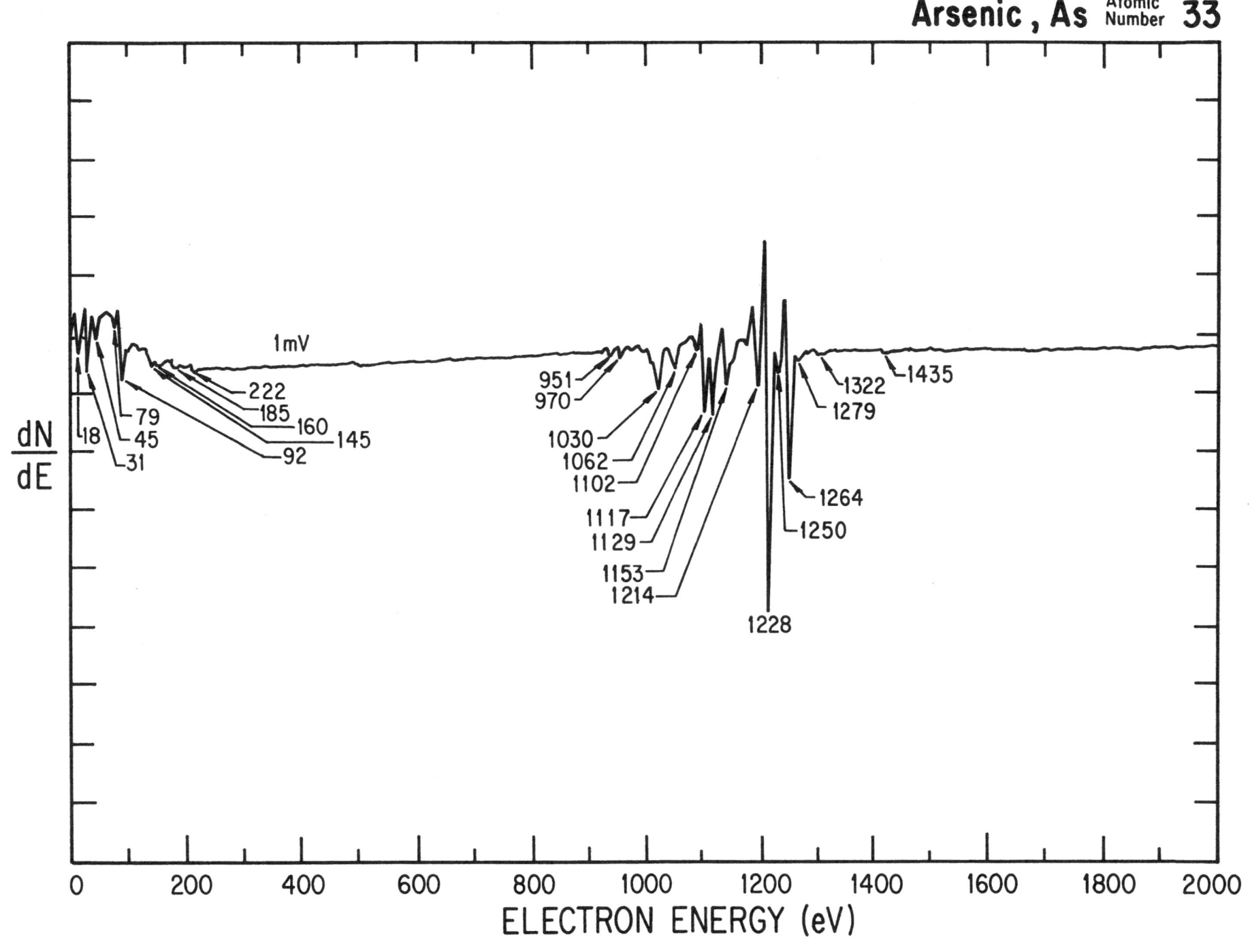

$\dfrac{dN}{dE}$

ELECTRON ENERGY (eV)

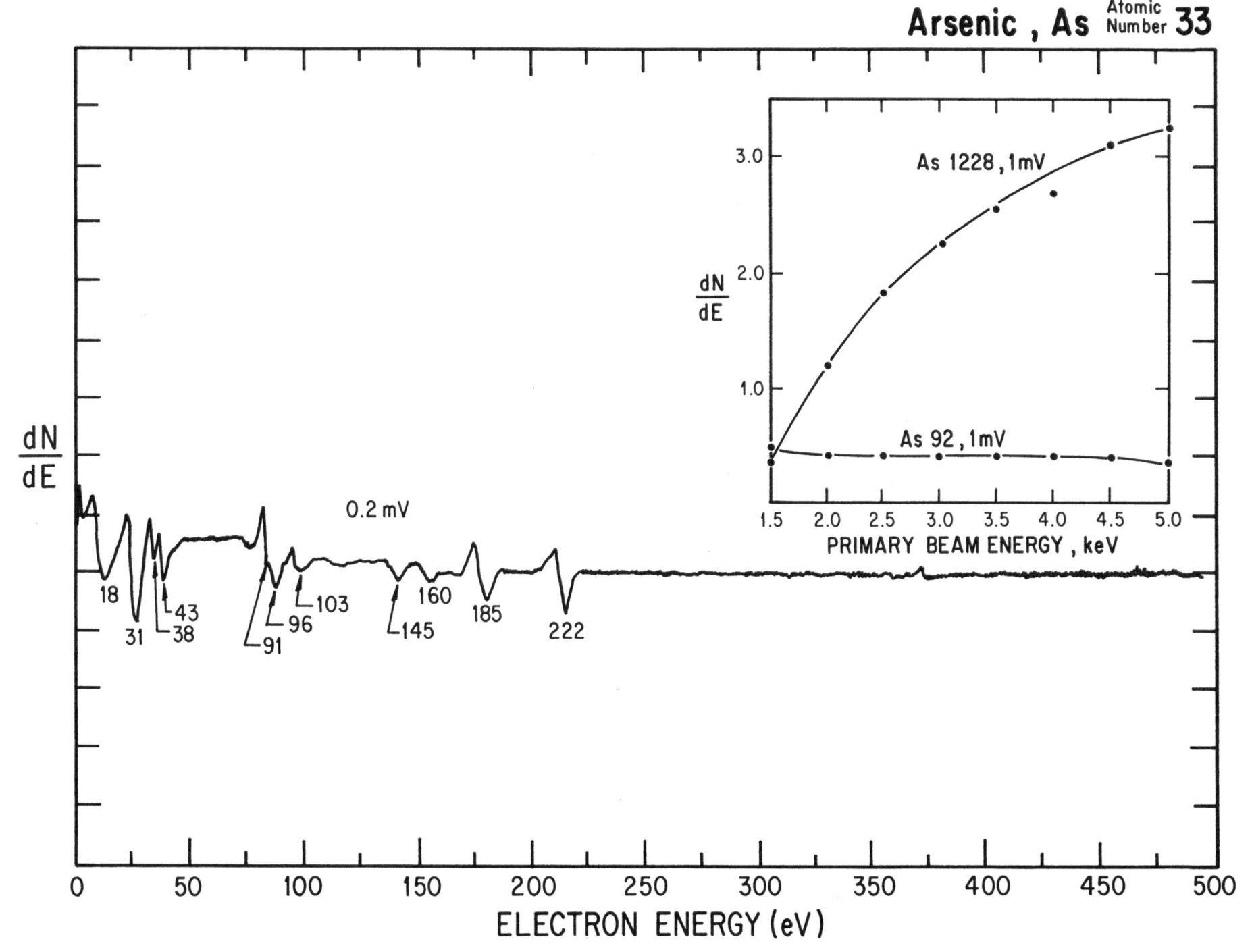

Arsenic , As Atomic Number 33

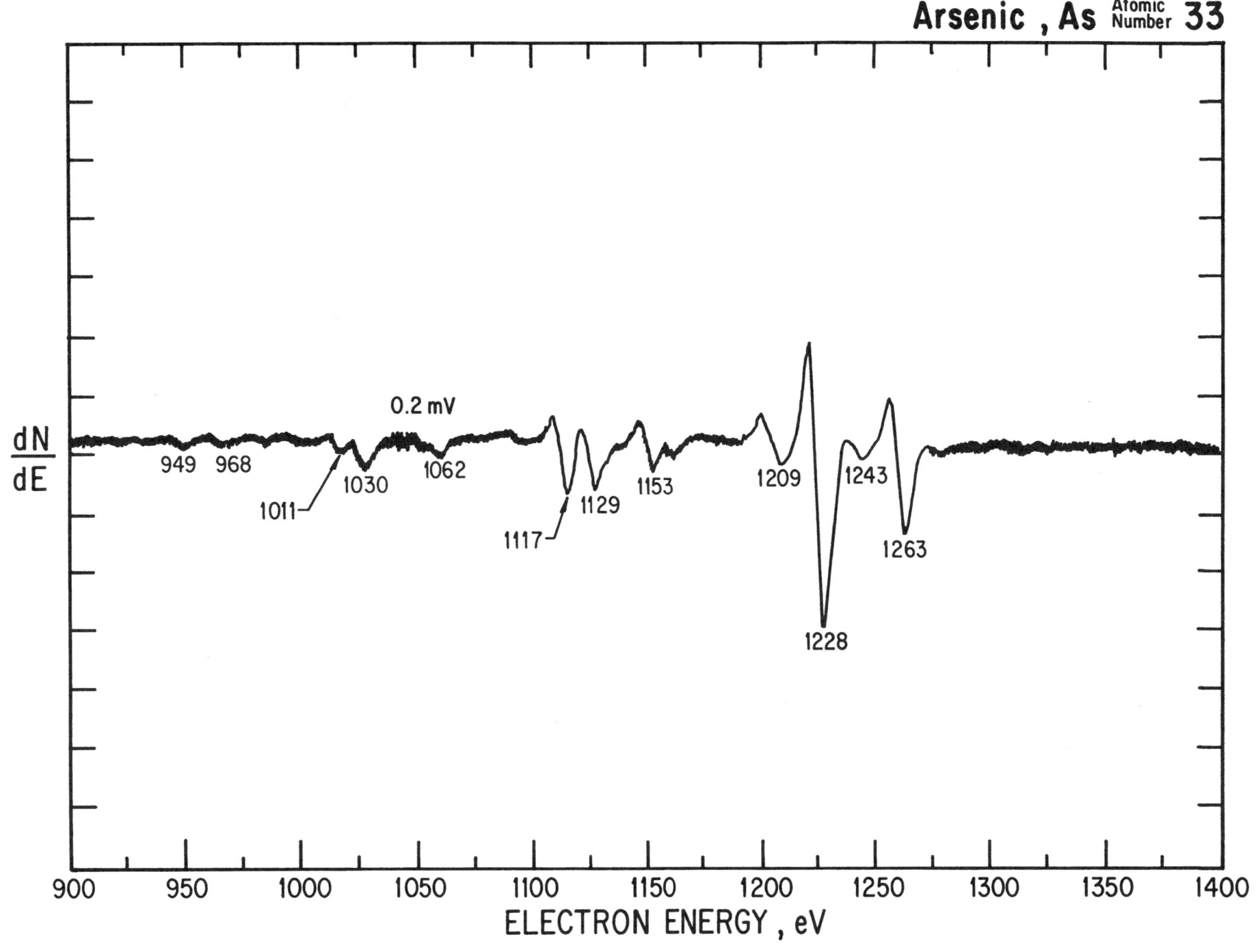

$\dfrac{dN}{dE}$

0.2 mV

949   968

1011

1030

1062

1117

1129

1153

1209

1243

1228

1263

ELECTRON ENERGY , eV

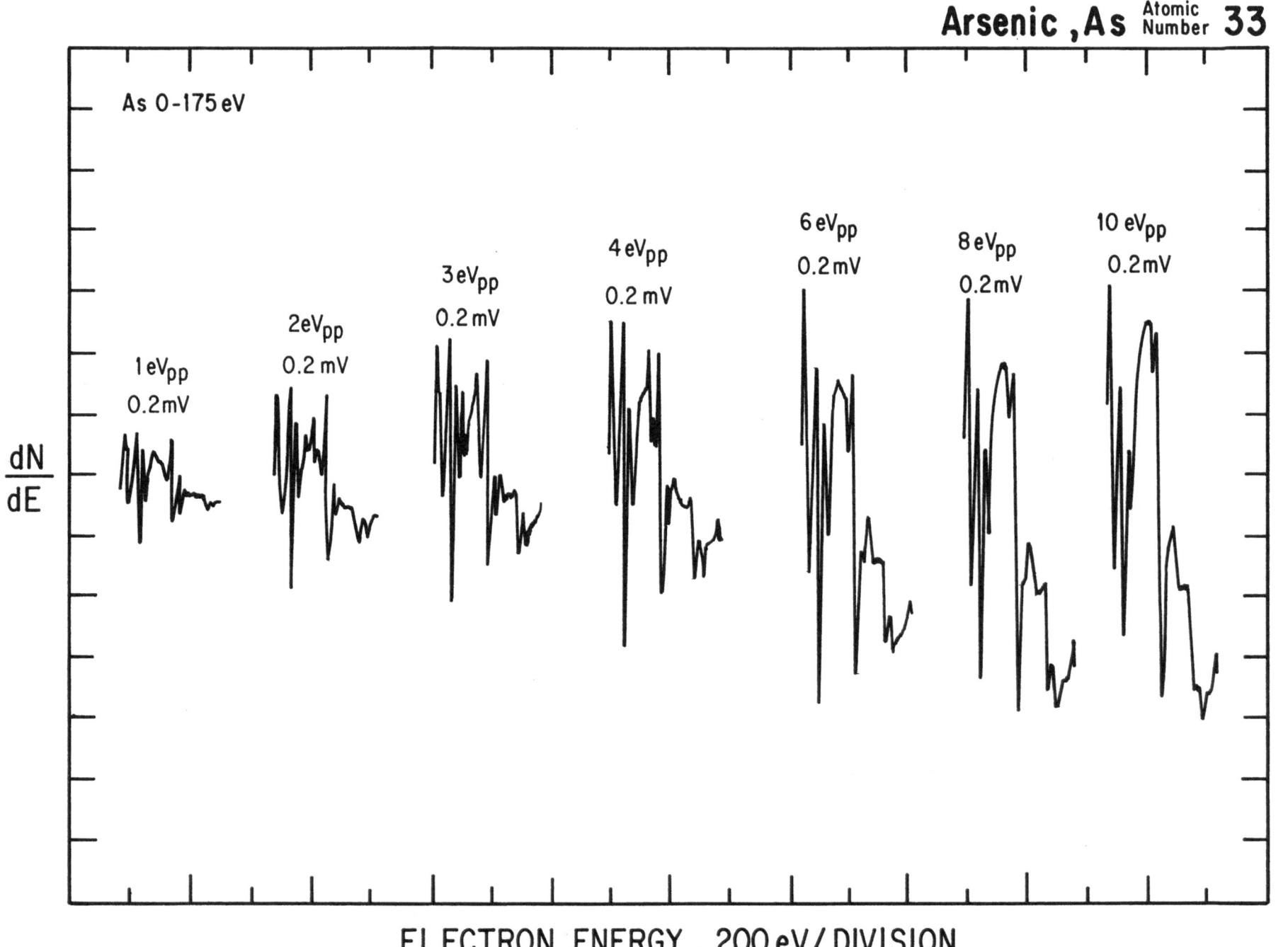

As 0-175 eV

1 eV_pp
0.2 mV

2 eV_pp
0.2 mV

3 eV_pp
0.2 mV

4 eV_pp
0.2 mV

6 eV_pp
0.2 mV

8 eV_pp
0.2 mV

10 eV_pp
0.2 mV

$\dfrac{dN}{dE}$

ELECTRON ENERGY , 200 eV / DIVISION

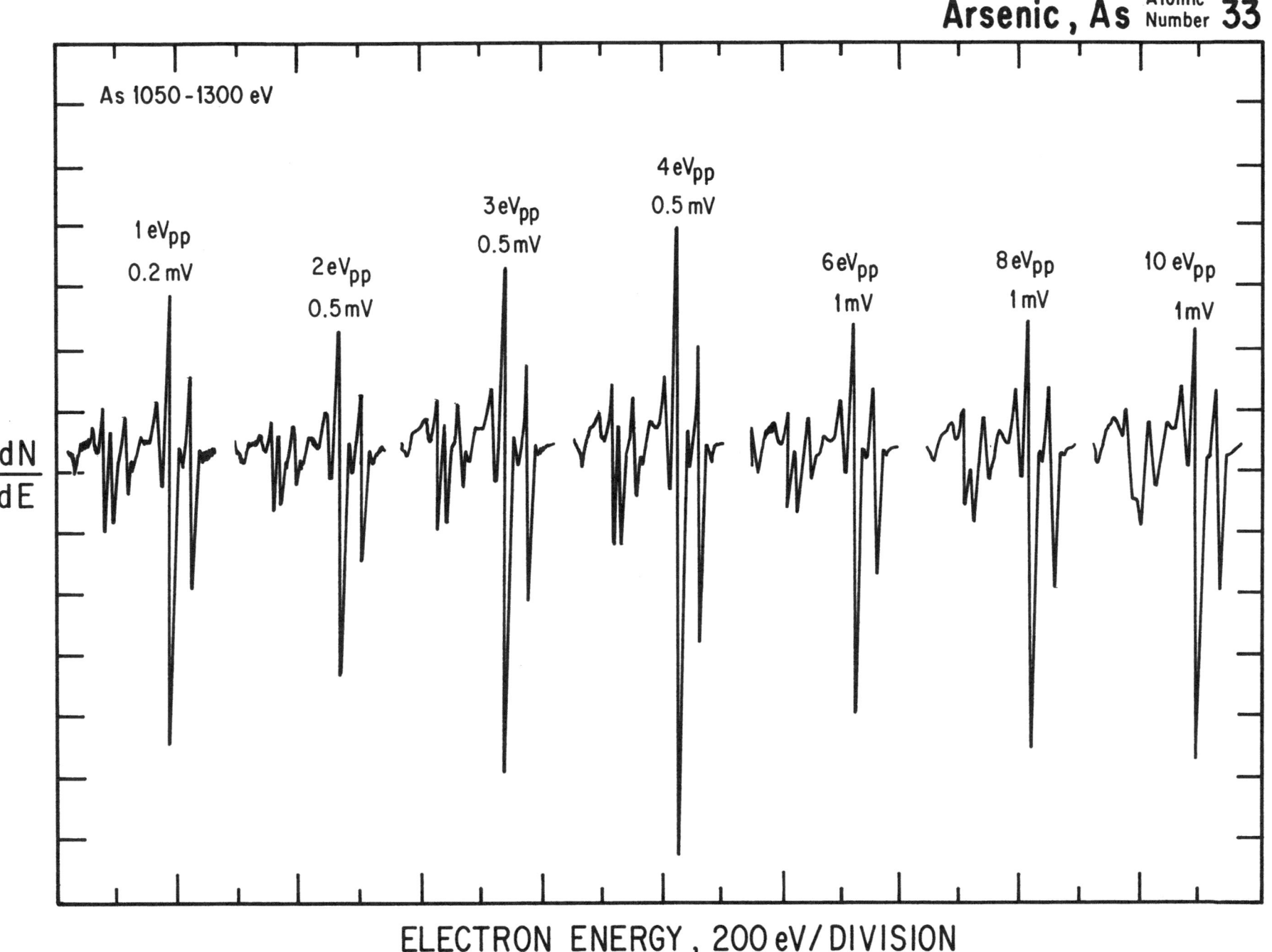

As 1050 - 1300 eV

1 eV$_{pp}$
0.2 mV

2 eV$_{pp}$
0.5 mV

3 eV$_{pp}$
0.5 mV

4 eV$_{pp}$
0.5 mV

6 eV$_{pp}$
1 mV

8 eV$_{pp}$
1 mV

10 eV$_{pp}$
1 mV

$\dfrac{dN}{dE}$

ELECTRON ENERGY, 200 eV / DIVISION

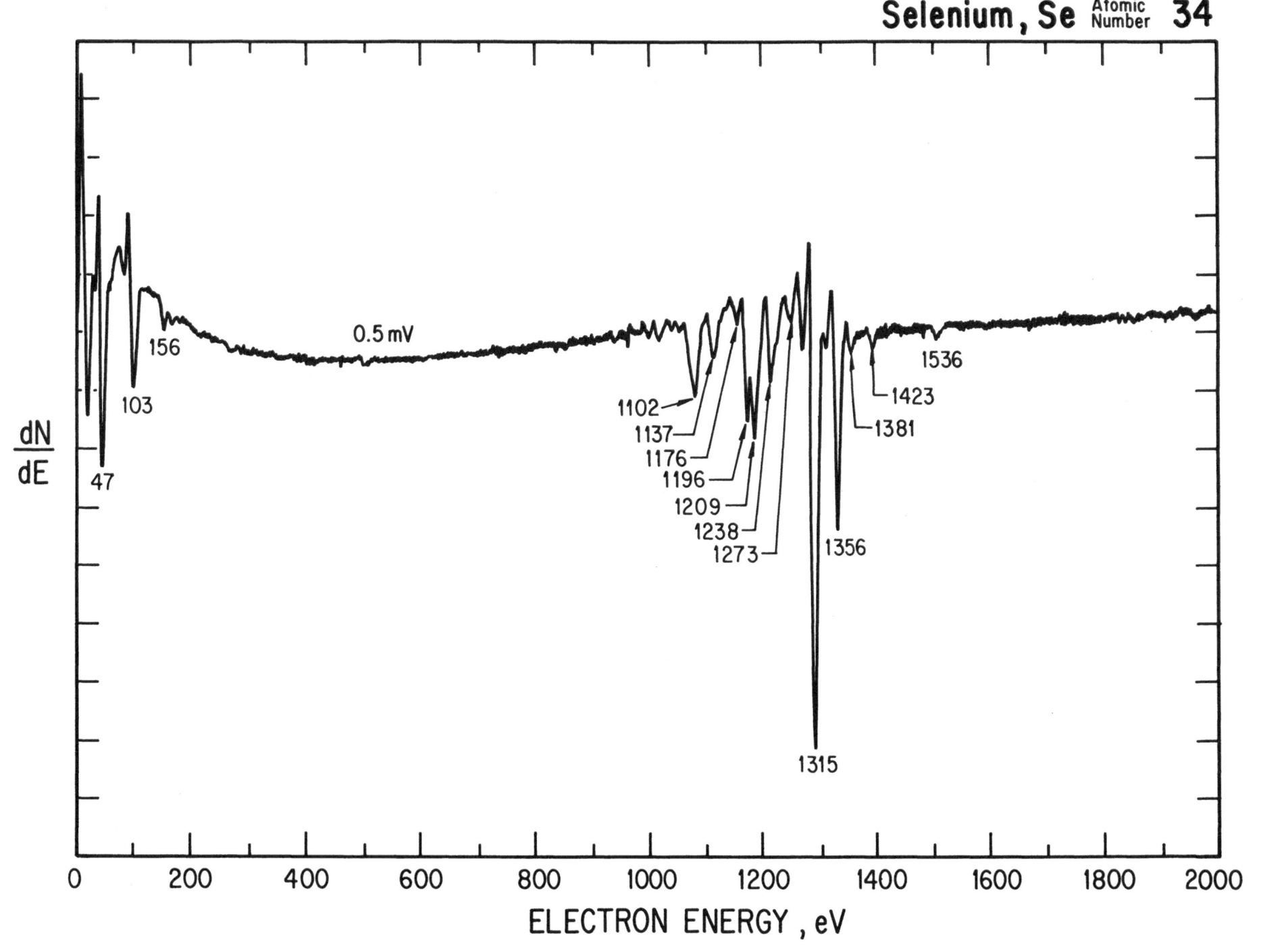

$\dfrac{dN}{dE}$

0.5 mV

47

103

156

1102
1137
1176
1196
1209
1238
1273

1315

1356

1381
1423

1536

ELECTRON ENERGY , eV

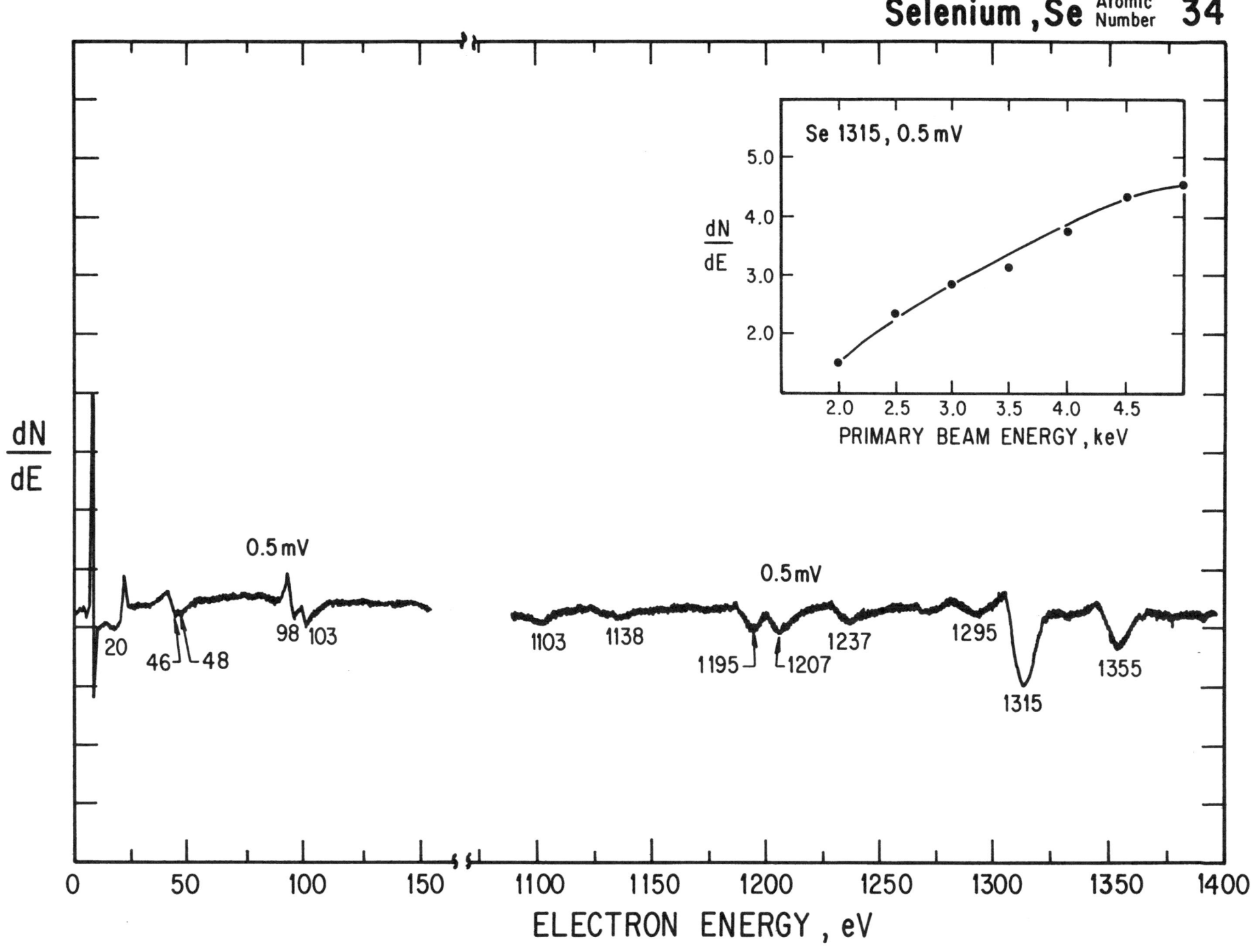

$\dfrac{dN}{dE}$

Se 1315, 0.5 mV

$\dfrac{dN}{dE}$

5.0

4.0

3.0

2.0

PRIMARY BEAM ENERGY , keV

0.5 mV

0.5 mV

20

46 48

98 103

1103 1138

1195 1207 1237

1295

1315

1355

ELECTRON ENERGY , eV

Selenium, Se Atomic Number **34**

Se 1225 - 1400 eV

1 eV_pp 0.5 mV

2 eV_pp 0.5 mV

3 eV_pp 0.5 mV

4 eV_pp 0.5 mV

6 eV_pp 0.5 mV

8 eV_pp 0.5 mV

10 eV_pp 0.5 mV

$\dfrac{dN}{dE}$

ELECTRON ENERGY, 200 eV / DIVISION

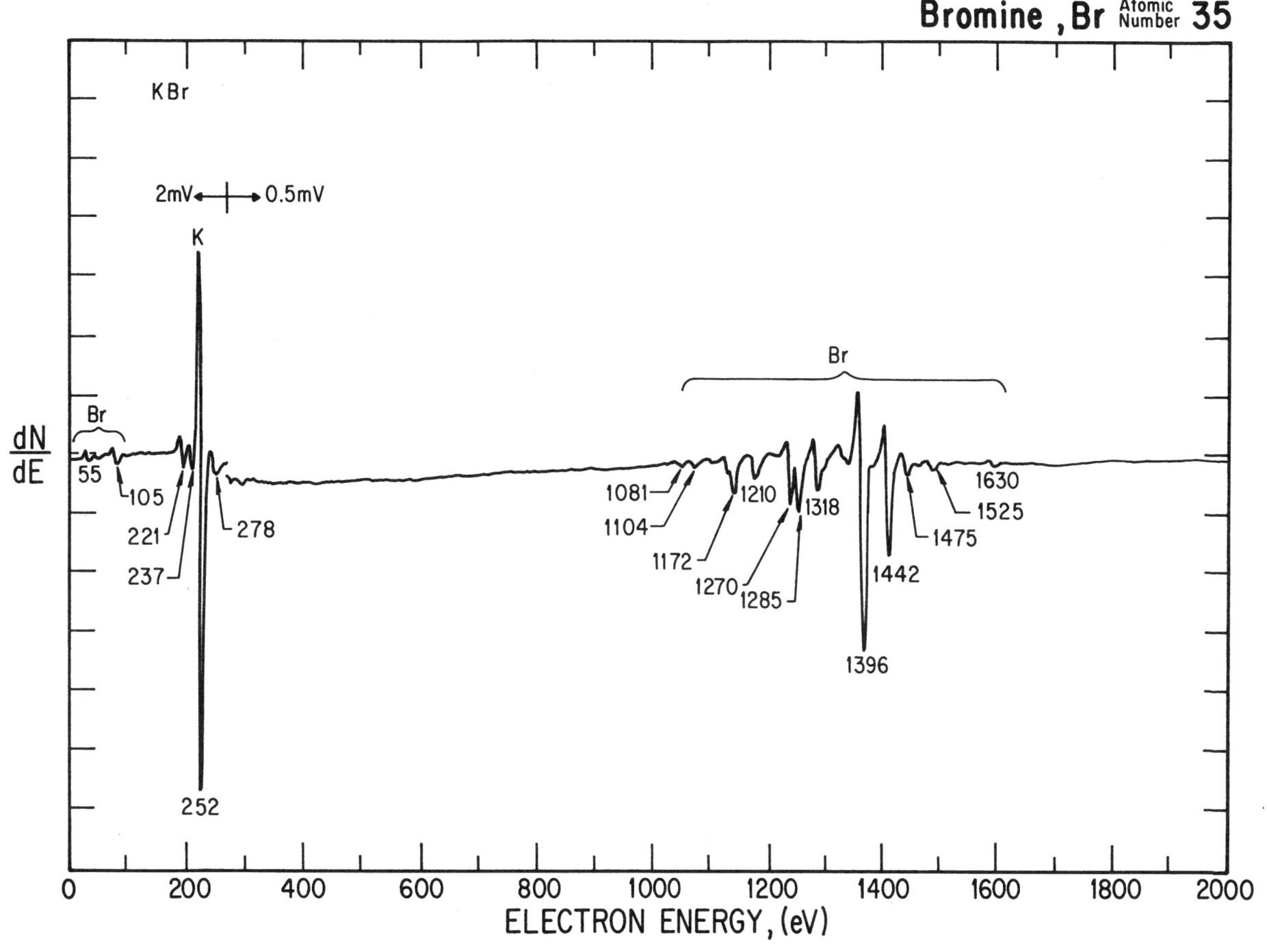

Bromine , Br Atomic Number 35

$\frac{dN}{dE}$

KBr

2mV ← | → 0.5mV

K

Br

Br

55

105

221

237

278

252

1081

1104

1172

1270

1285

1210

1318

1396

1442

1475

1525

1630

ELECTRON ENERGY, (eV)

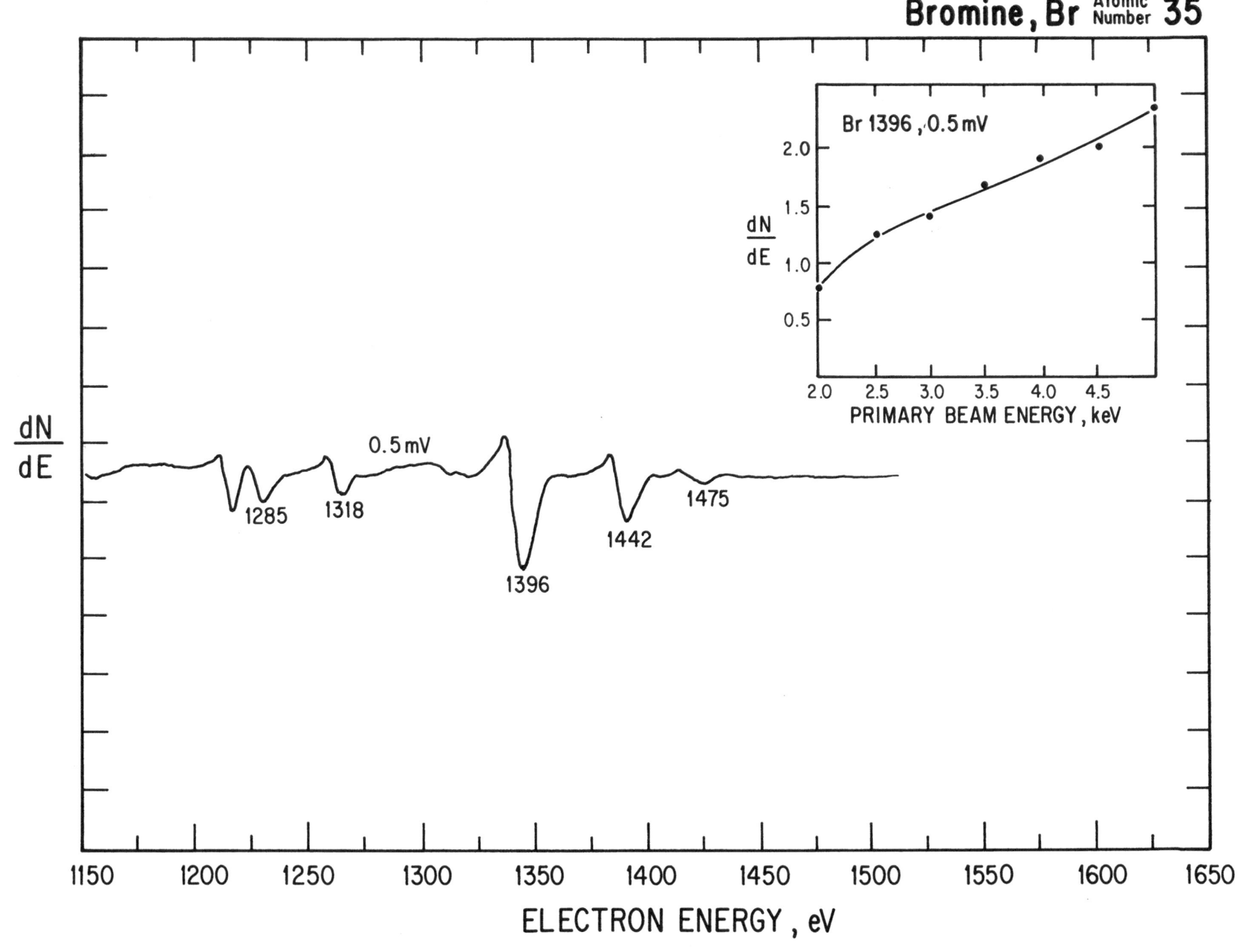

Bromine, Br Atomic Number 35

Br 1396 , 0.5 mV

0.5 mV

1285

1318

1396

1442

1475

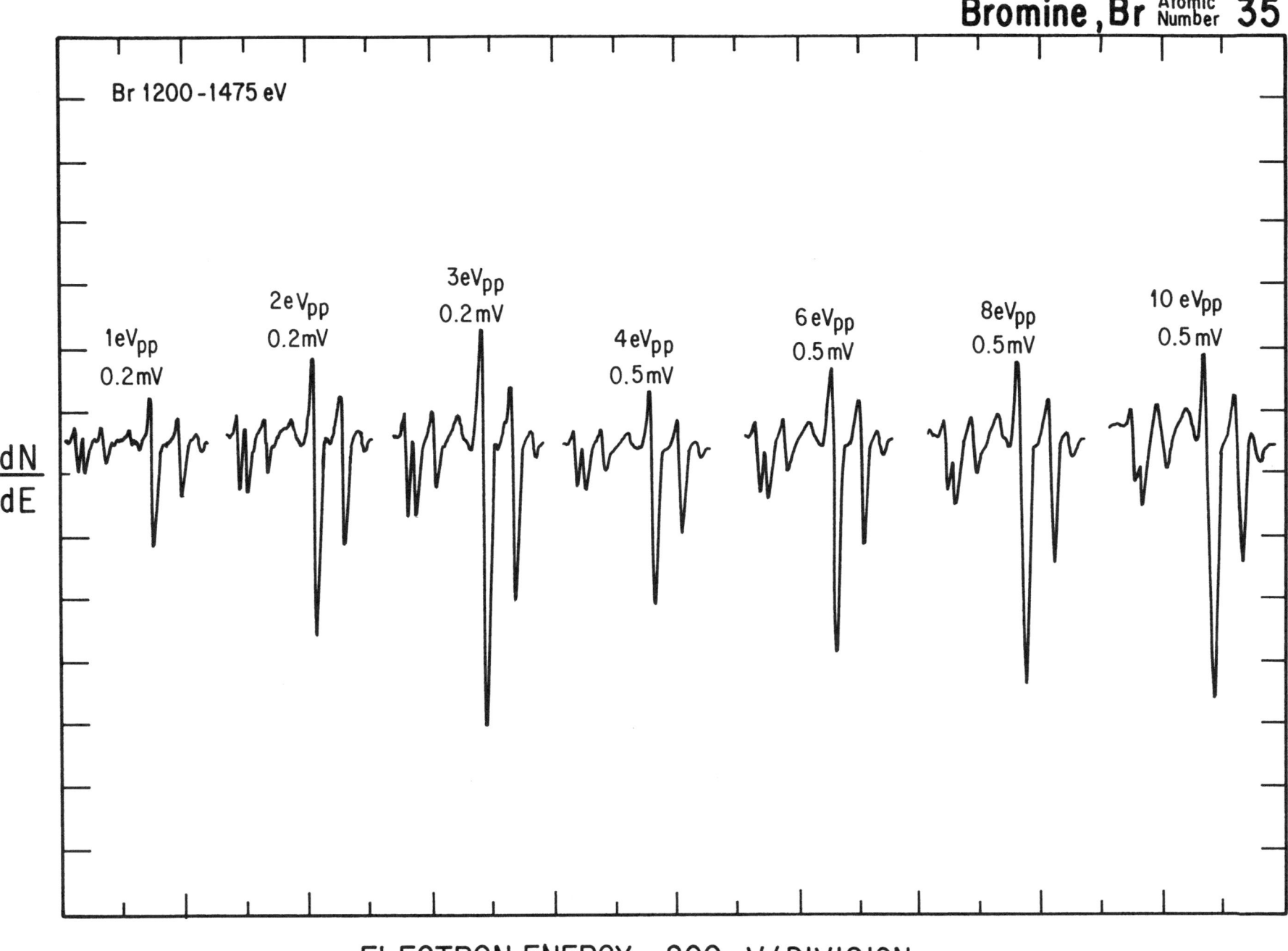

Br 1200 - 1475 eV

1eV_pp 0.2mV

2eV_pp 0.2mV

3eV_pp 0.2mV

4eV_pp 0.5mV

6eV_pp 0.5mV

8eV_pp 0.5mV

10 eV_pp 0.5 mV

$\dfrac{dN}{dE}$

ELECTRON ENERGY , 200 eV/DIVISION

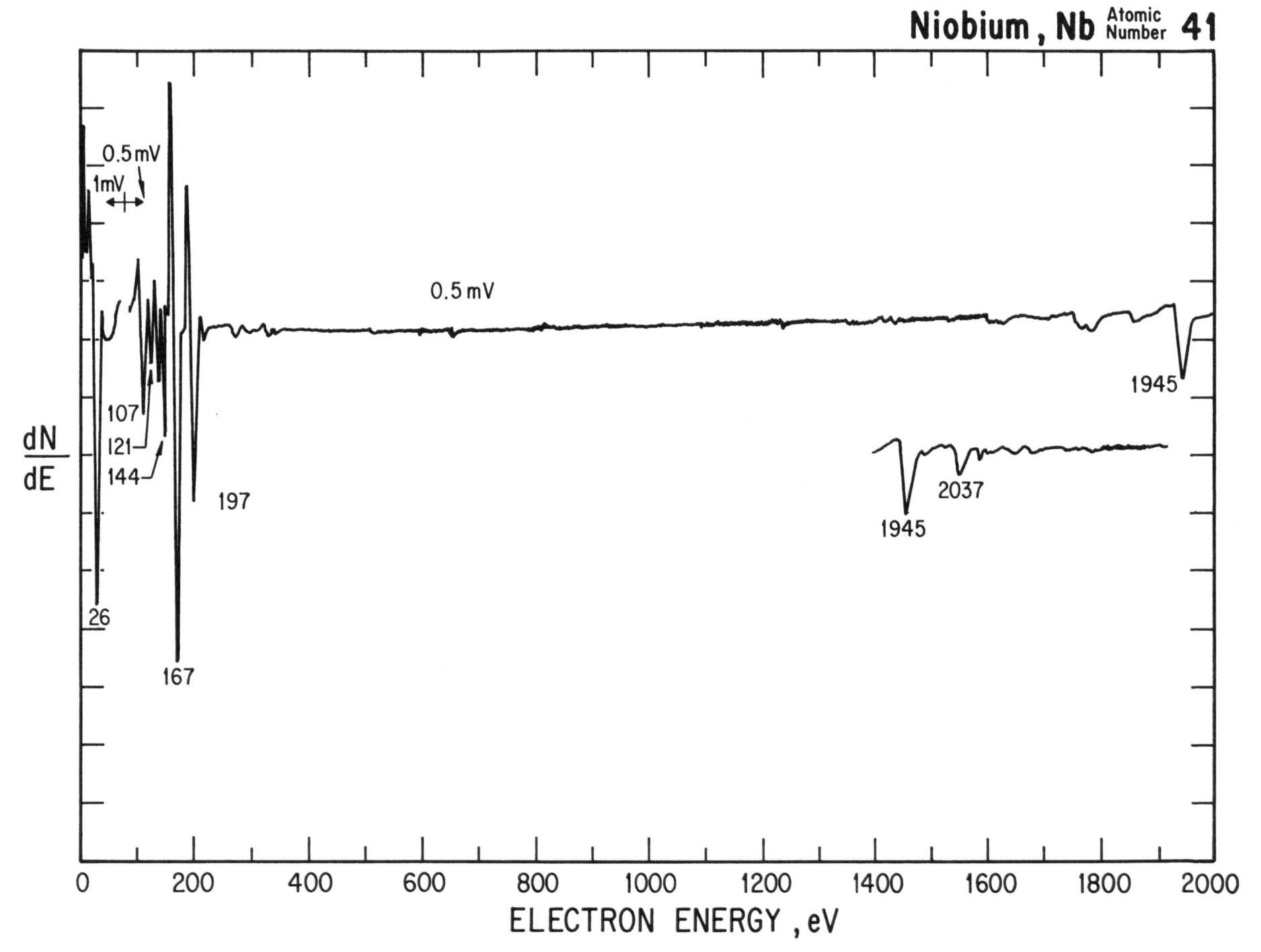

0.5 mV

1 mV

0.5 mV

1945

107

121

144

197

26

167

1945

2037

dN / dE

ELECTRON ENERGY , eV

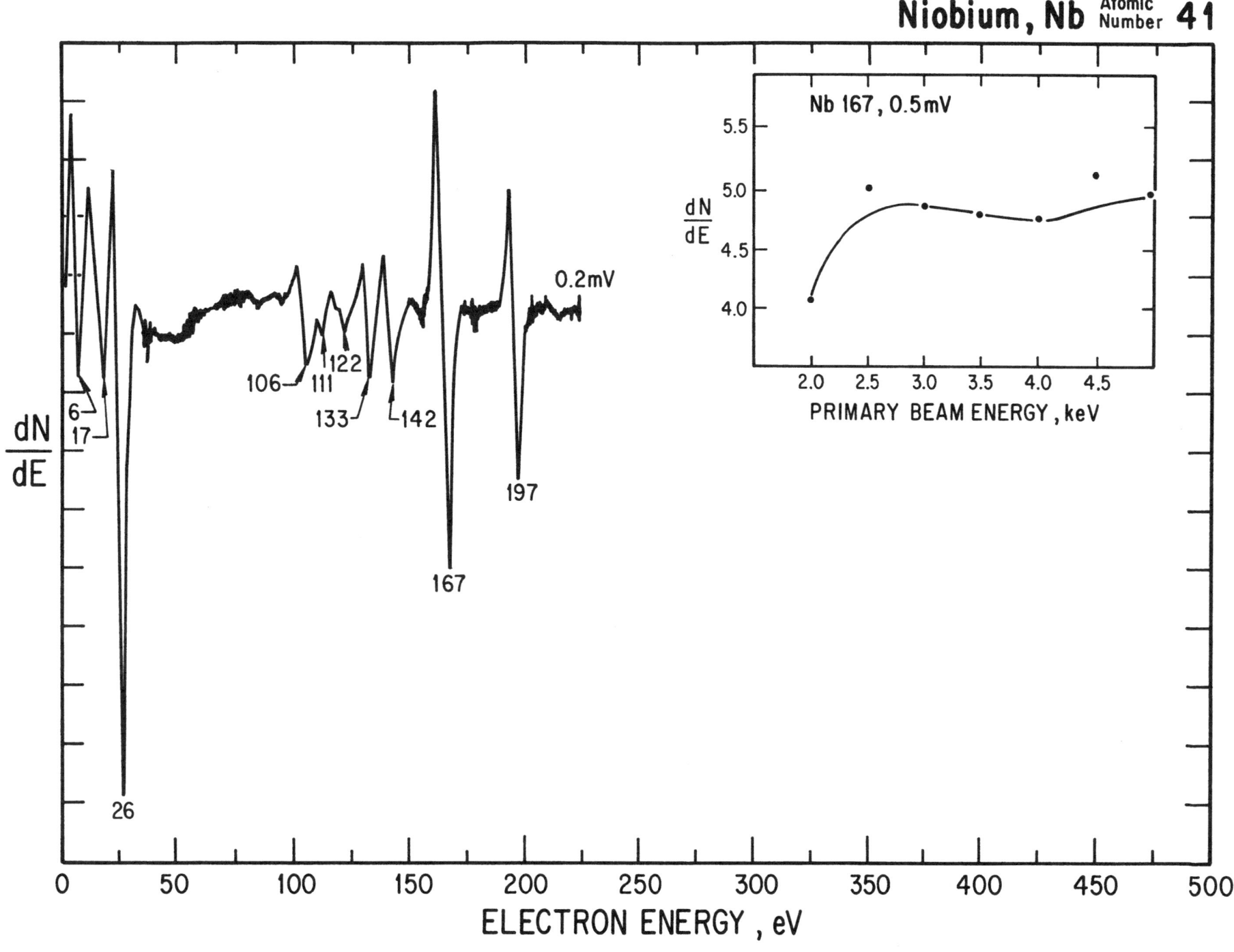

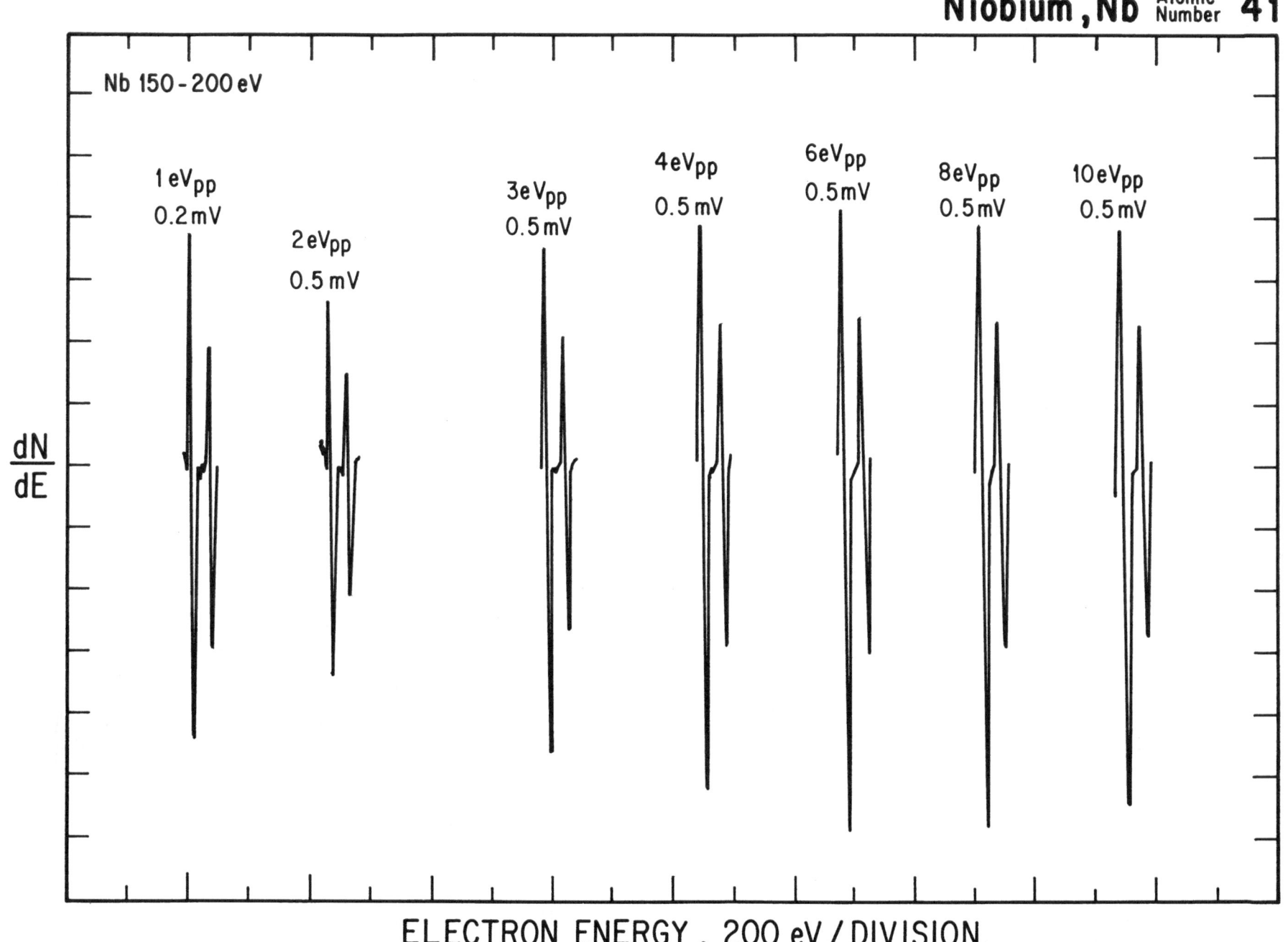

Nb 150-200 eV

1 eV$_{pp}$
0.2 mV

2 eV$_{pp}$
0.5 mV

3 eV$_{pp}$
0.5 mV

4 eV$_{pp}$
0.5 mV

6 eV$_{pp}$
0.5 mV

8 eV$_{pp}$
0.5 mV

10 eV$_{pp}$
0.5 mV

$\dfrac{dN}{dE}$

ELECTRON ENERGY , 200 eV / DIVISION

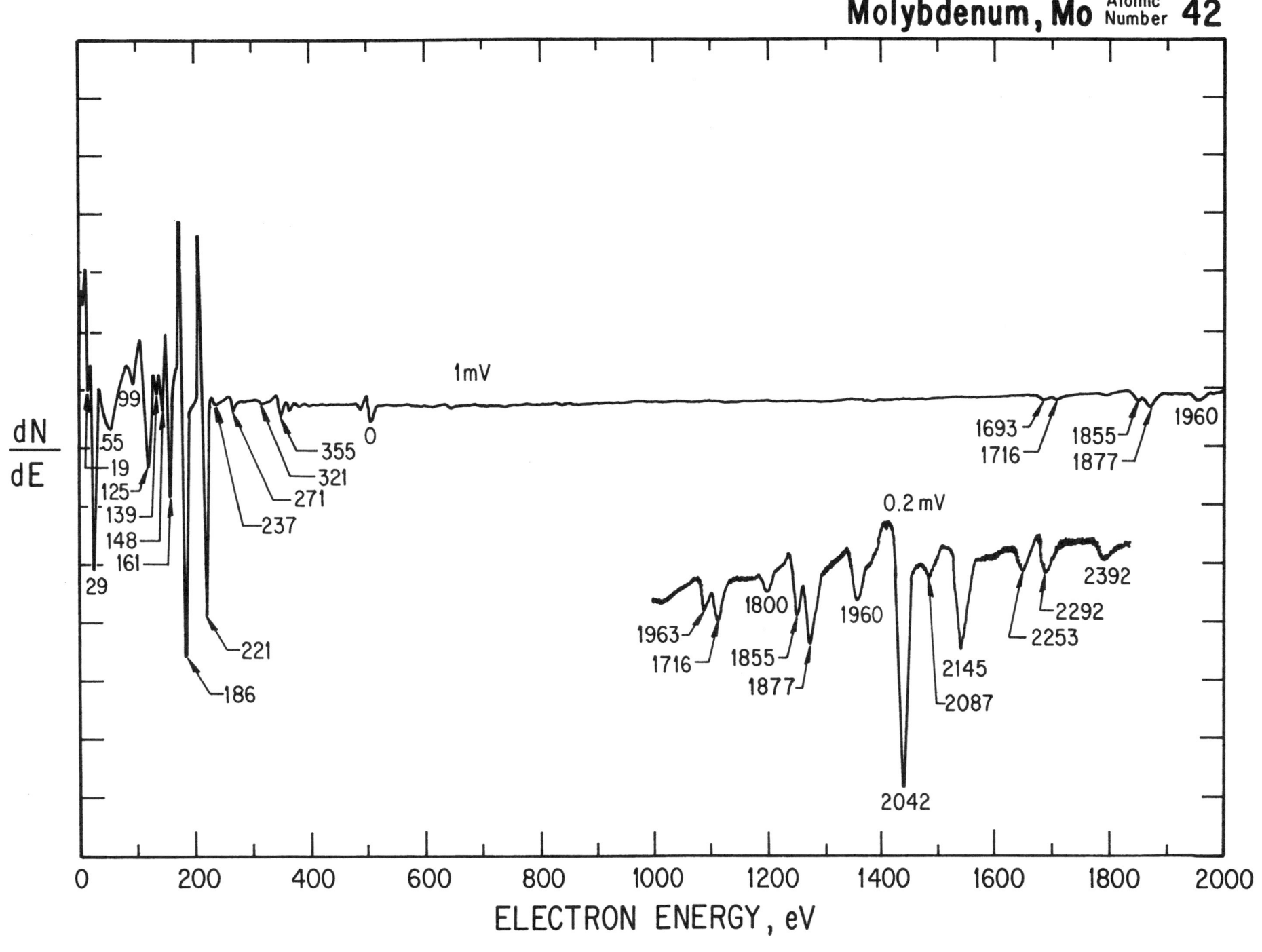

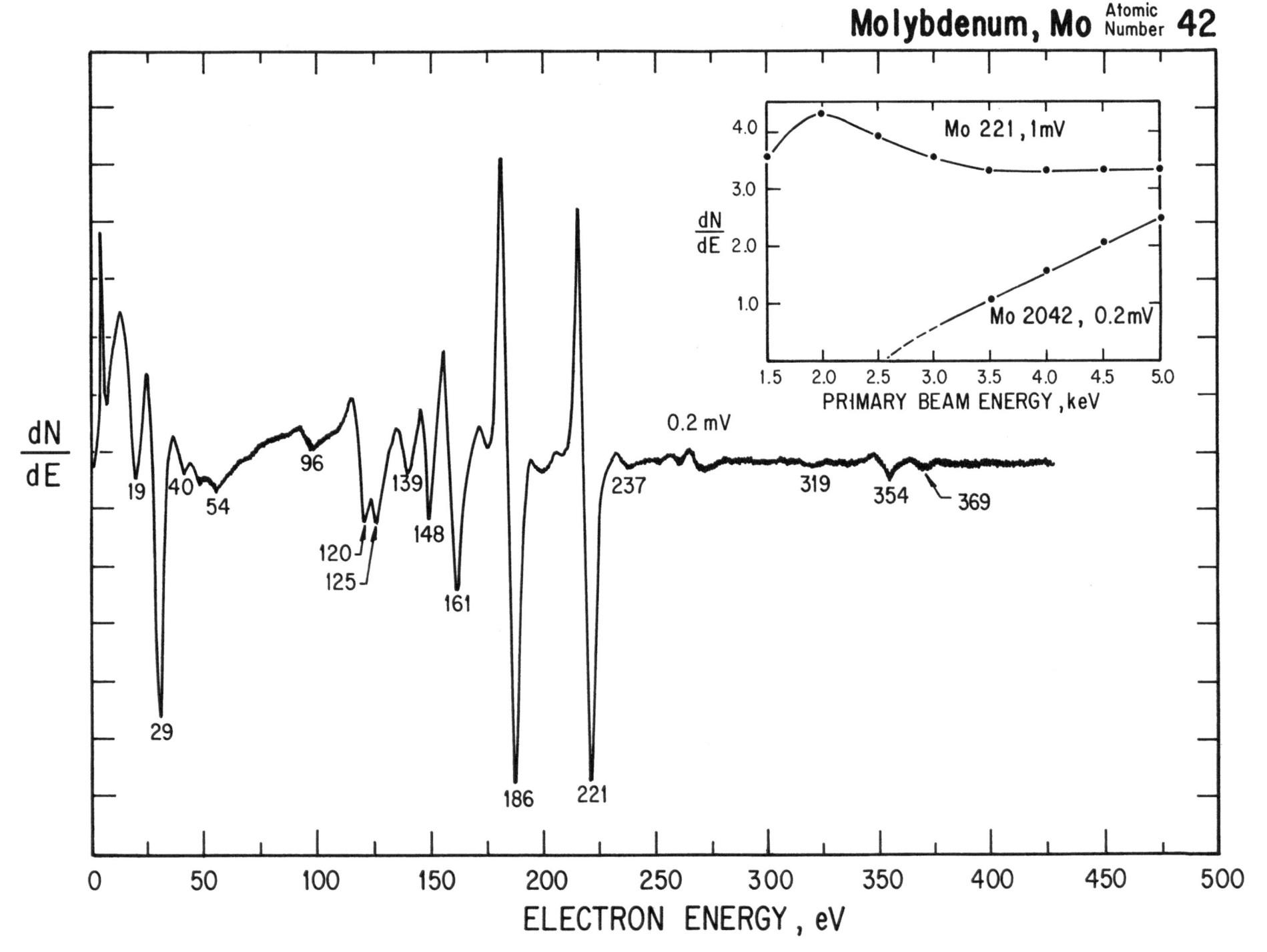

Molybdenum, Mo Atomic Number 42

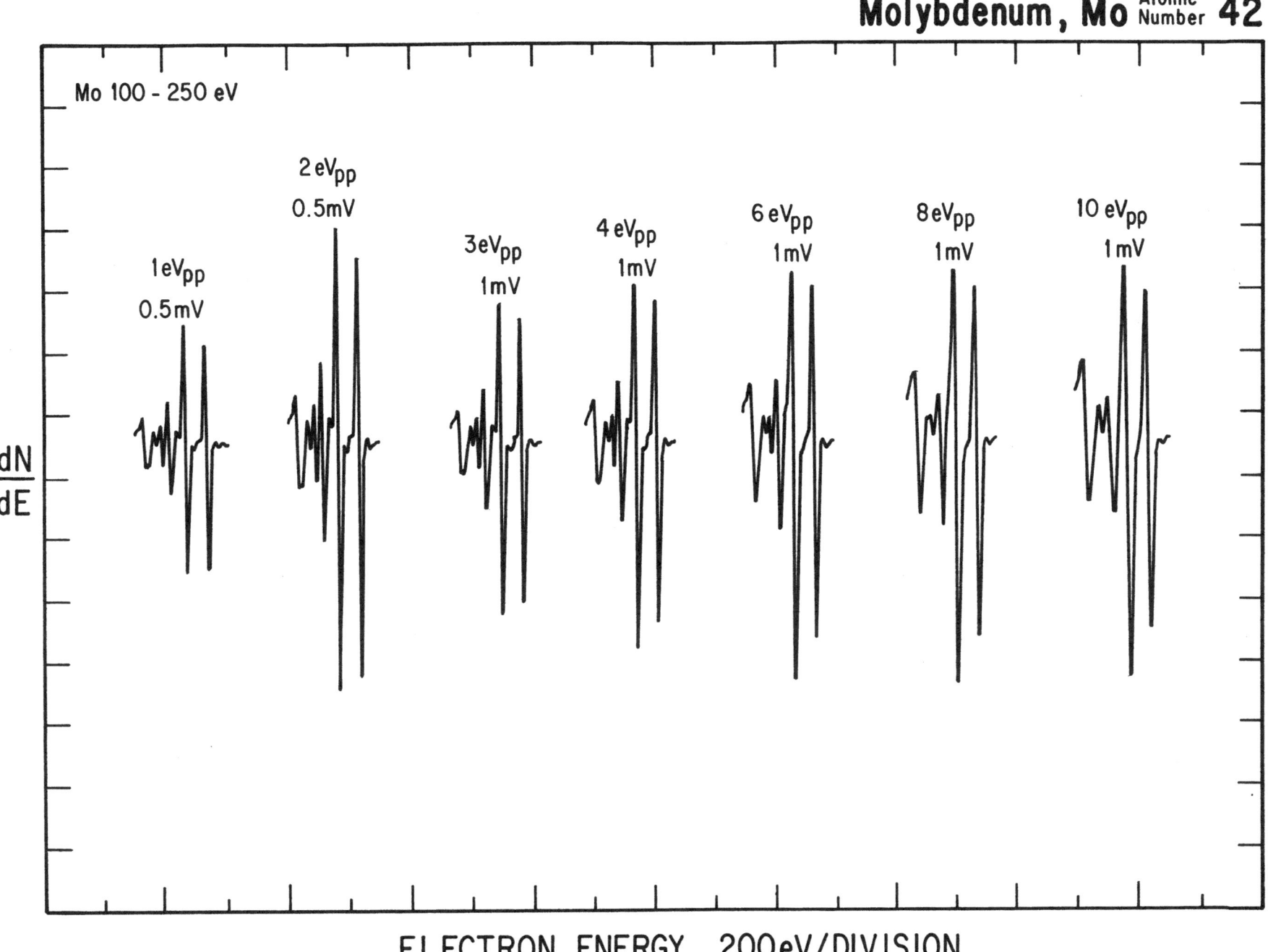

Mo 100 - 250 eV

1eV$_{pp}$
0.5mV

2 eV$_{pp}$
0.5mV

3eV$_{pp}$
1mV

4 eV$_{pp}$
1mV

6 eV$_{pp}$
1mV

8 eV$_{pp}$
1mV

10 eV$_{pp}$
1mV

$\dfrac{dN}{dE}$

ELECTRON ENERGY, 200eV/DIVISION

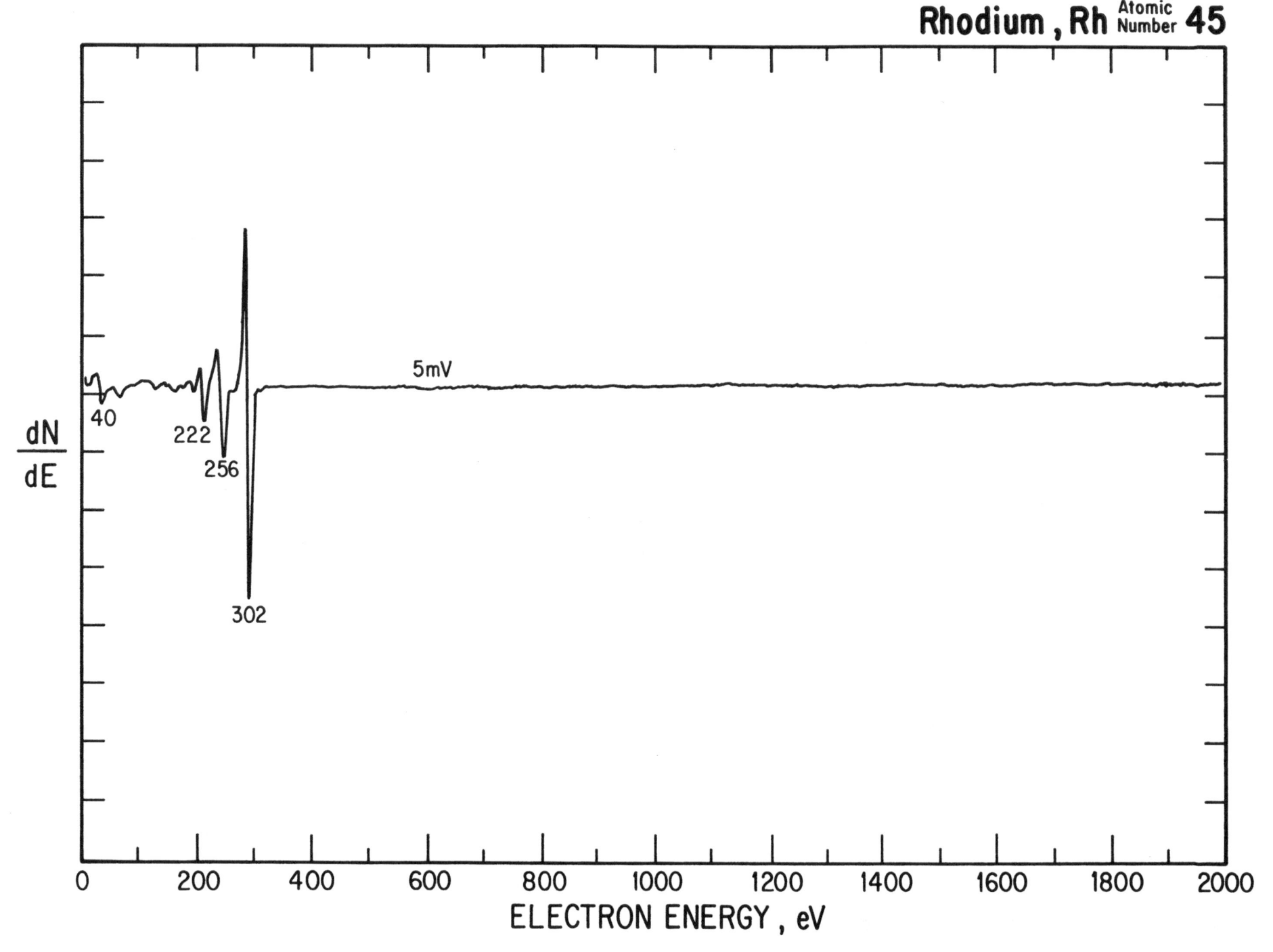

$\dfrac{dN}{dE}$

40

222

256

302

5mV

ELECTRON ENERGY , eV

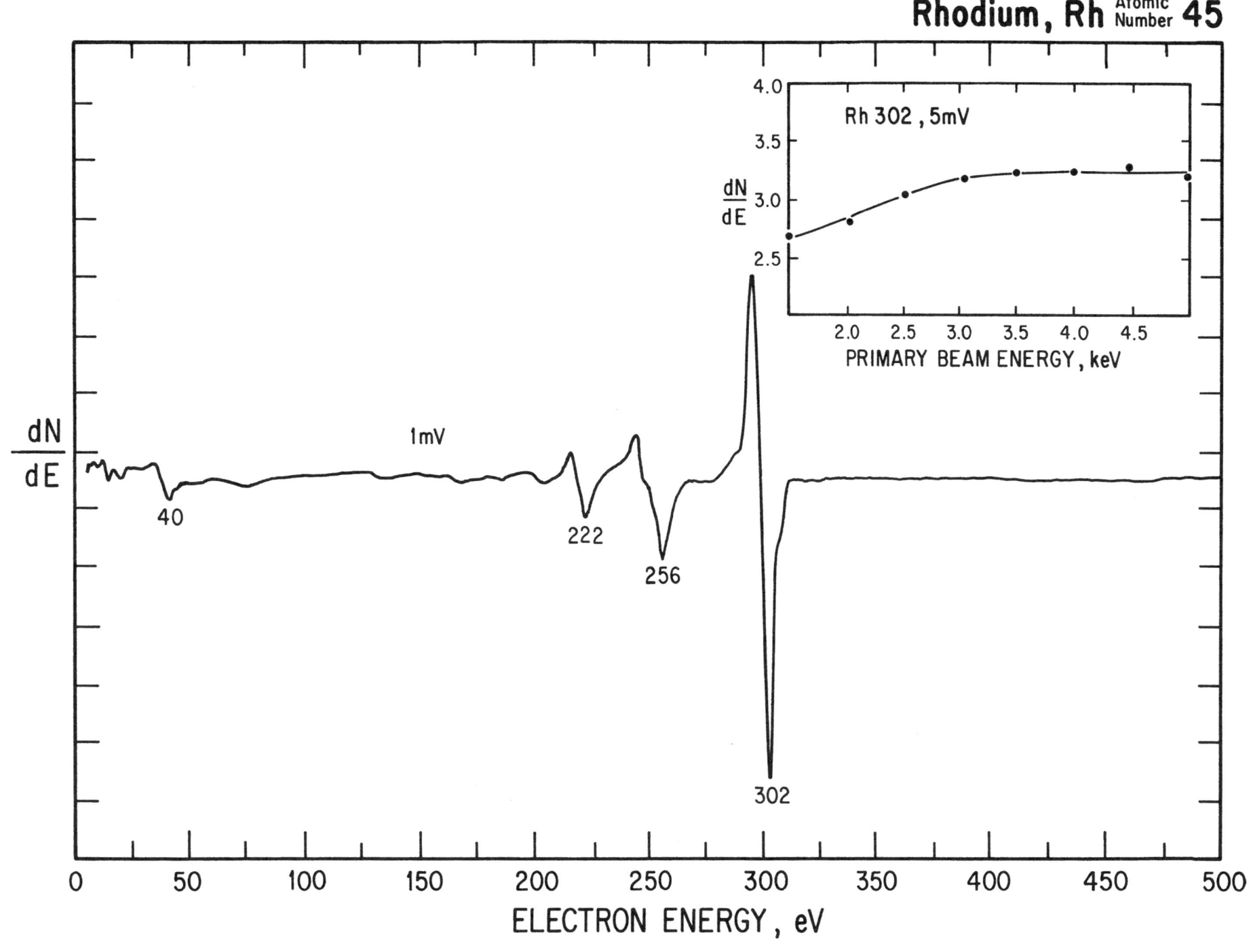

Rhodium, Rh Atomic Number **45**

Rh 302 , 5mV

$\dfrac{dN}{dE}$

PRIMARY BEAM ENERGY, keV

$\dfrac{dN}{dE}$

1mV

40

222

256

302

ELECTRON ENERGY, eV

Rhodium, Rh Atomic Number **45**

Rh 175 - 350 eV

1eV$_{pp}$ 1mV

2eV$_{pp}$ 2mV

3eV$_{pp}$ 2mV

4eV$_{pp}$ 5mV

6eV$_{pp}$ 5mV

8eV$_{pp}$ 5mV

10eV$_{pp}$ 5mV

$\dfrac{dN}{dE}$

ELECTRON ENERGY, 200 eV/DIVISION

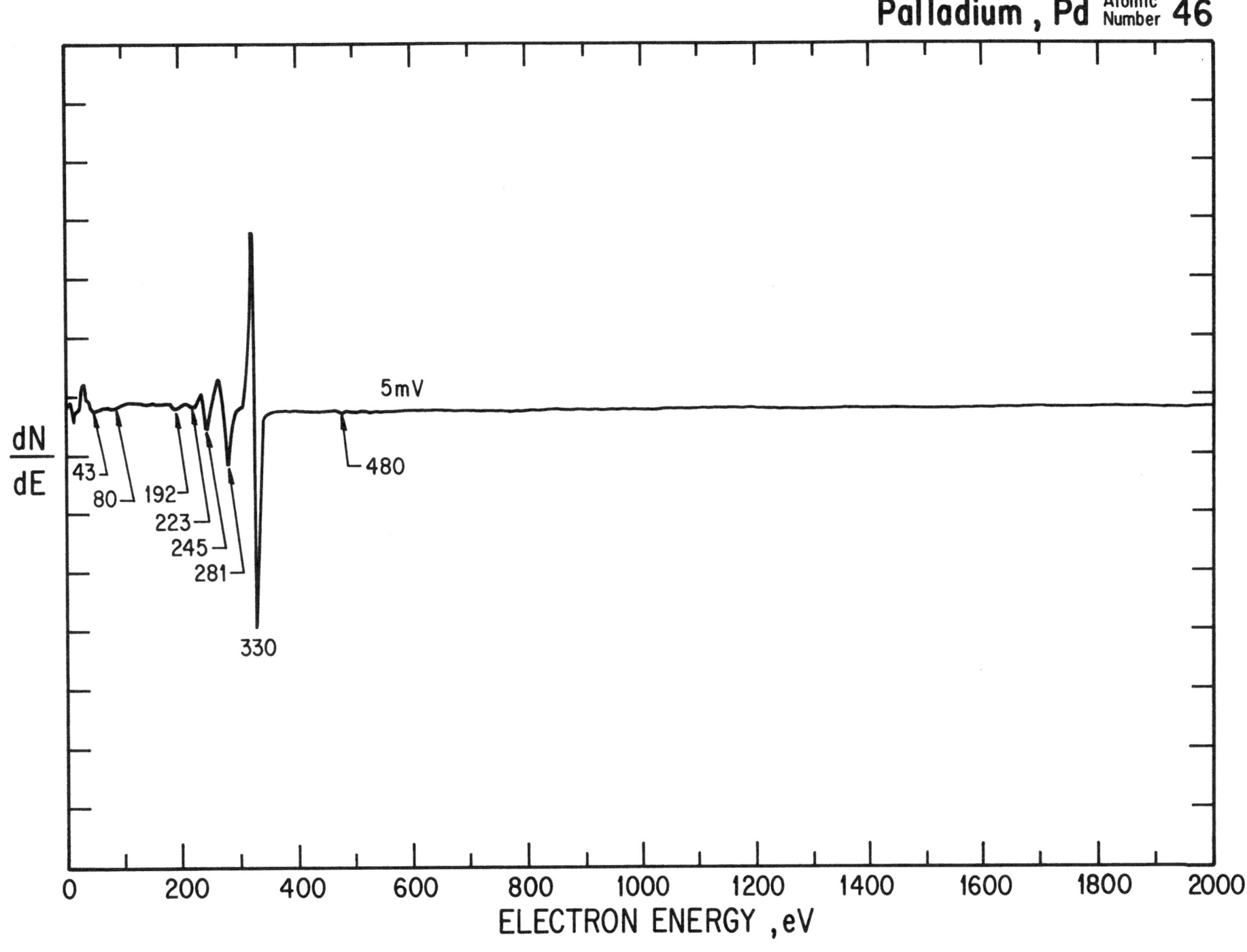

$\dfrac{dN}{dE}$

5mV

43
80
192
223
245
281
330
480

ELECTRON ENERGY , eV

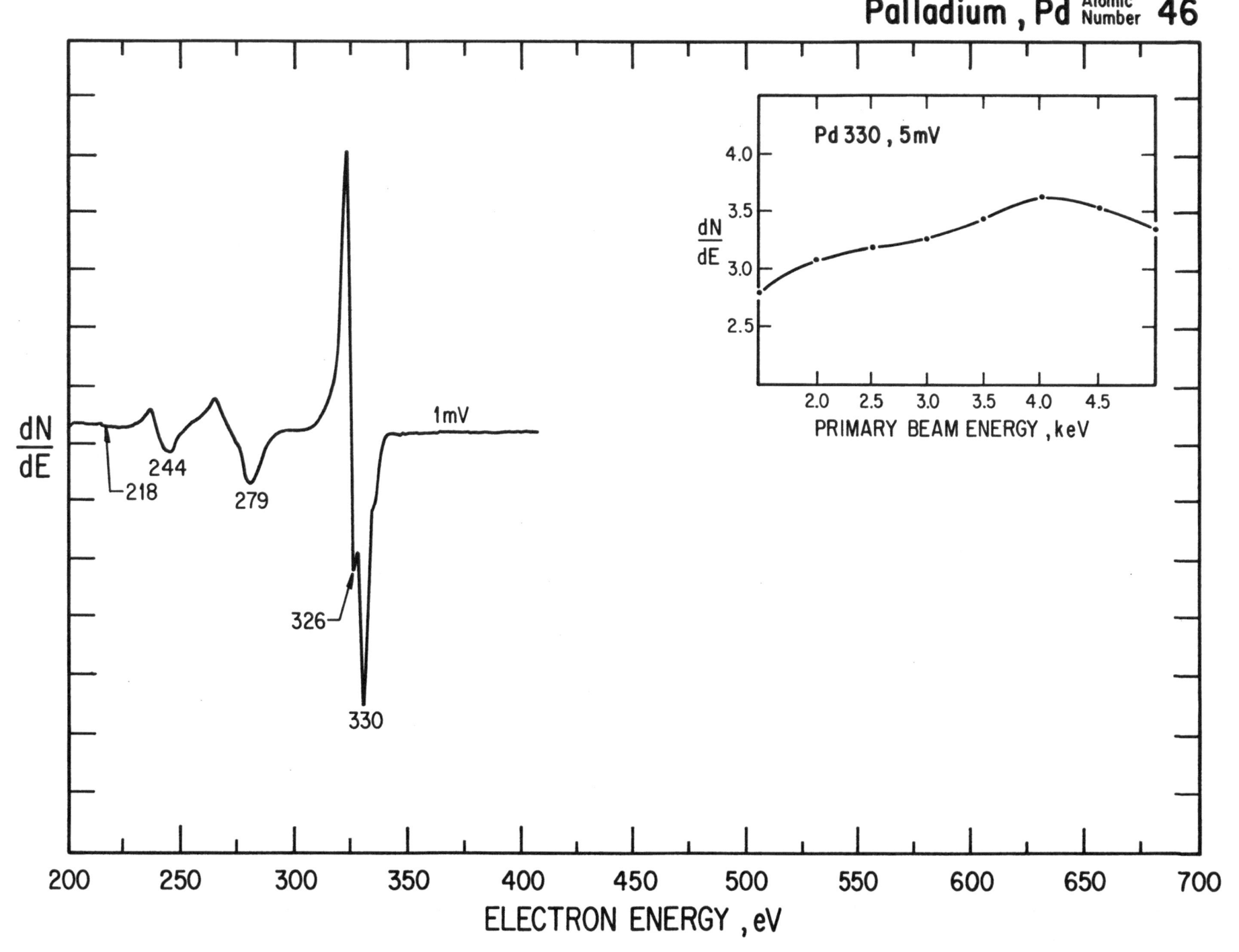

ELECTRON ENERGY , eV

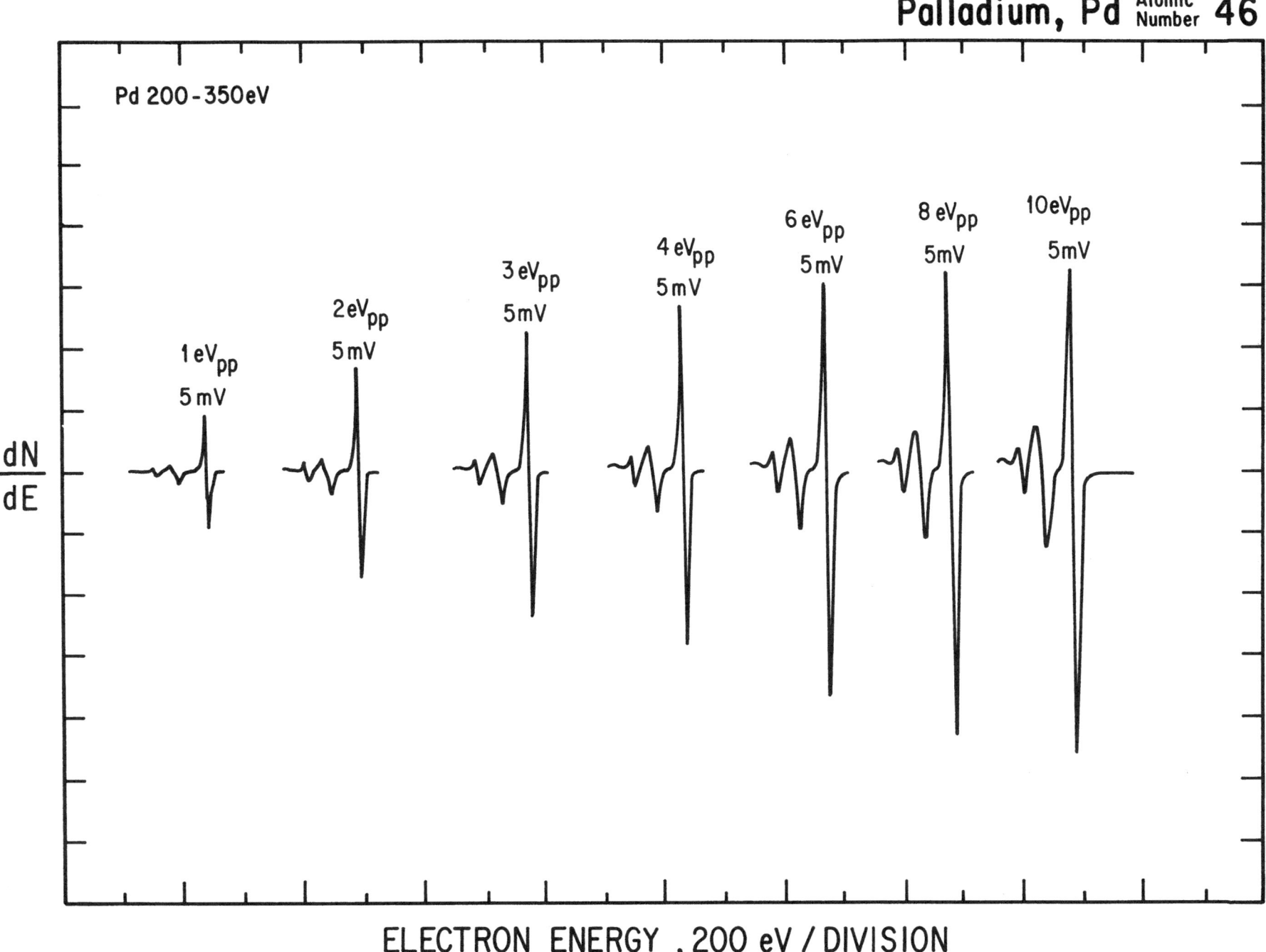

Pd 200-350eV

1 eV$_{pp}$
5 mV

2 eV$_{pp}$
5 mV

3 eV$_{pp}$
5 mV

4 eV$_{pp}$
5 mV

6 eV$_{pp}$
5 mV

8 eV$_{pp}$
5 mV

10 eV$_{pp}$
5 mV

$\dfrac{dN}{dE}$

ELECTRON ENERGY , 200 eV / DIVISION

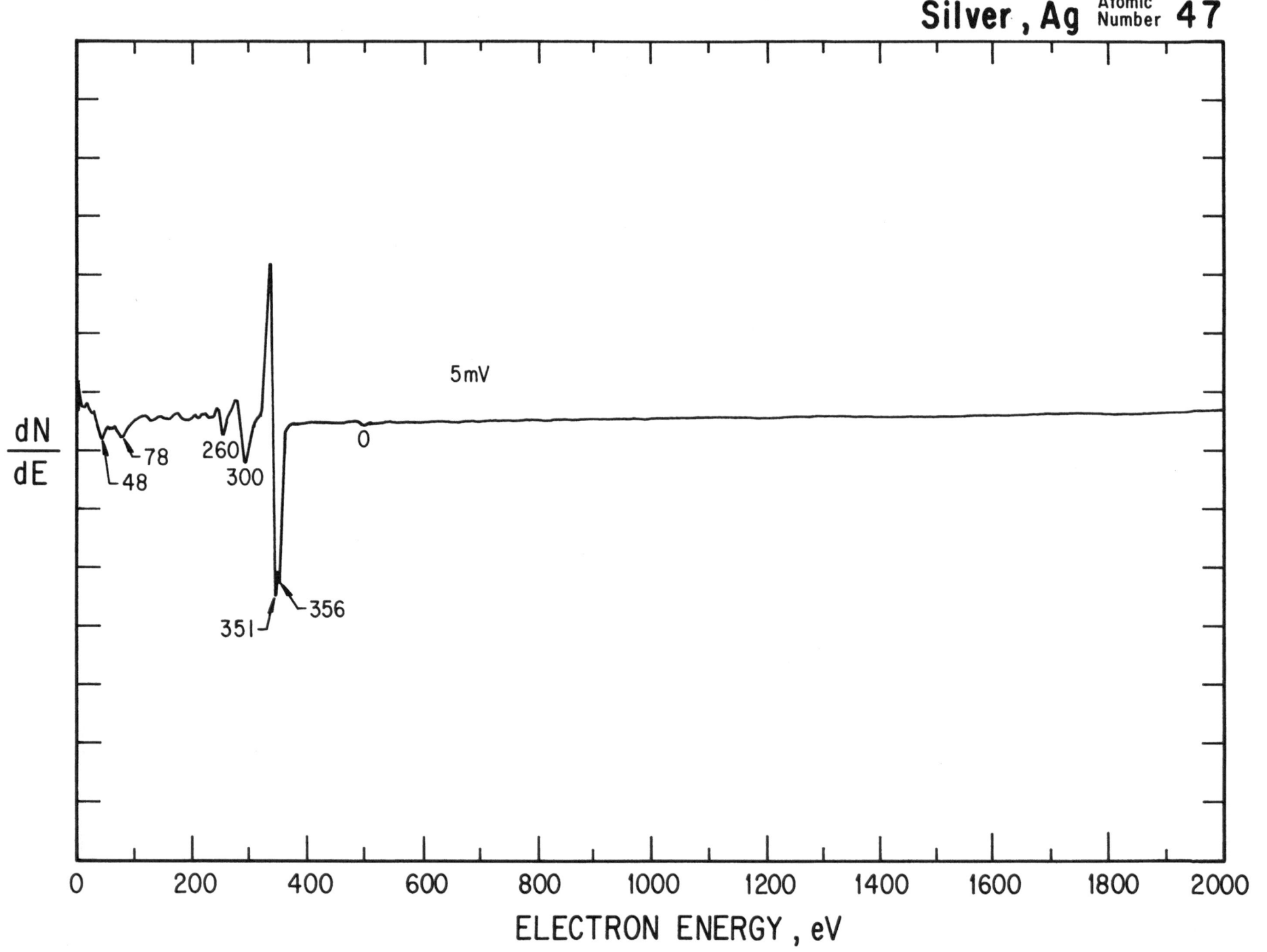

$\dfrac{dN}{dE}$

5mV

78
48
260
300
0
351
356

ELECTRON ENERGY, eV

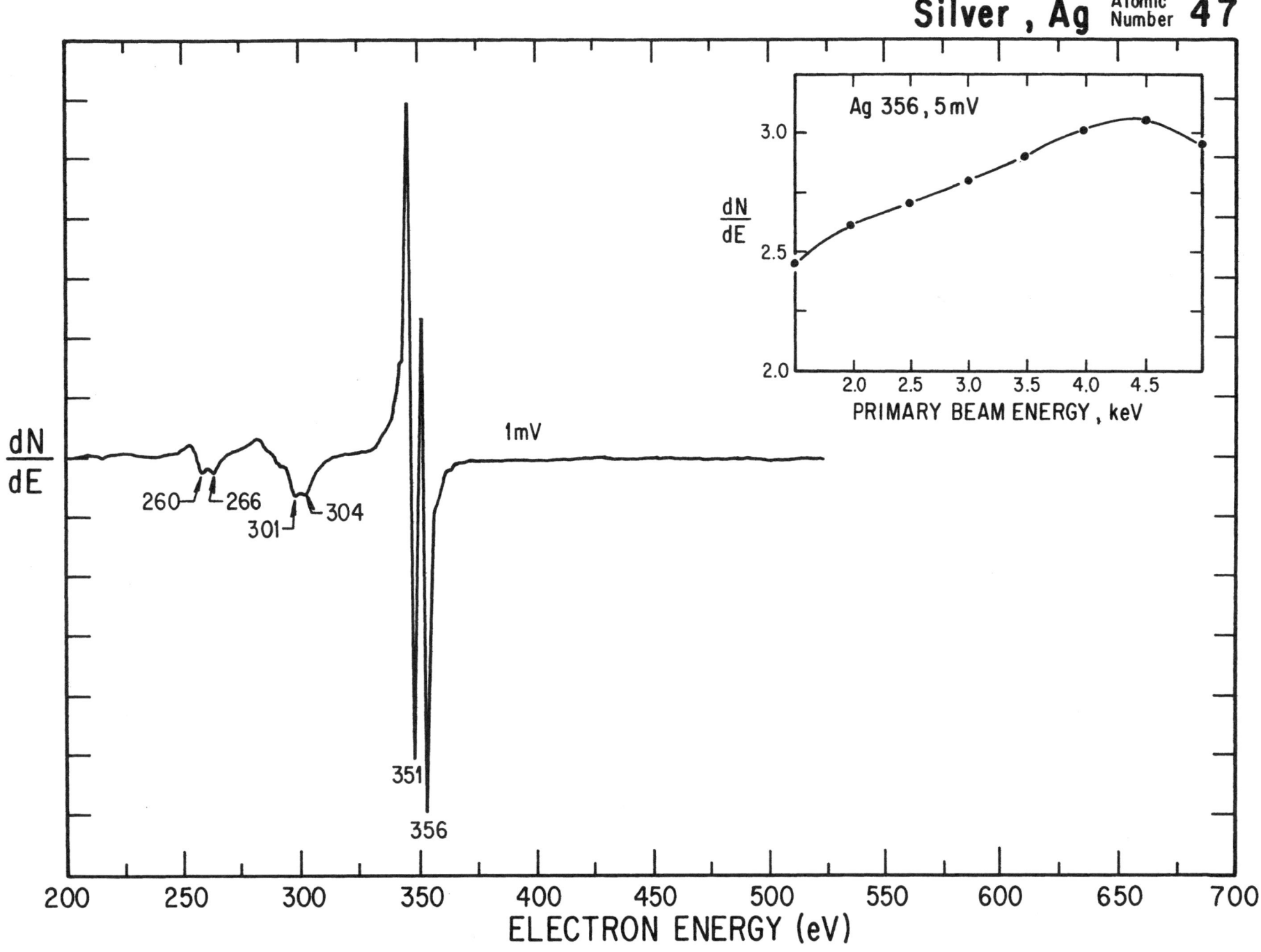

$\dfrac{dN}{dE}$

260 — 266
301 — 304

1mV

351

356

Ag 356 , 5mV

$\dfrac{dN}{dE}$

3.0

2.5

2.0

2.0 2.5 3.0 3.5 4.0 4.5

PRIMARY BEAM ENERGY, keV

200 250 300 350 400 450 500 550 600 650 700

ELECTRON ENERGY (eV)

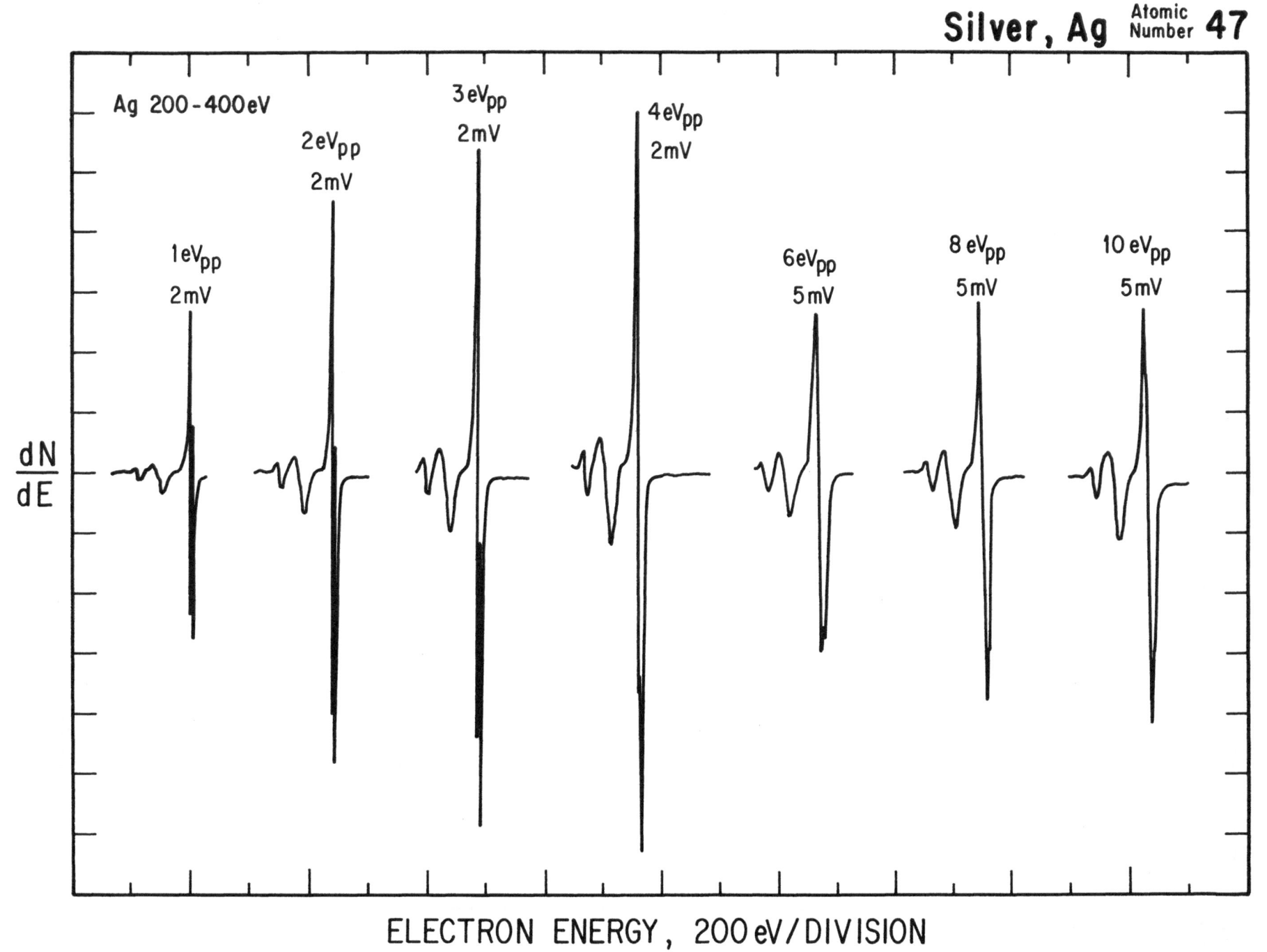

Ag 200-400eV

1eV$_{pp}$
2mV

2eV$_{pp}$
2mV

3eV$_{pp}$
2mV

4eV$_{pp}$
2mV

6eV$_{pp}$
5mV

8eV$_{pp}$
5mV

10eV$_{pp}$
5mV

$\dfrac{dN}{dE}$

ELECTRON ENERGY, 200eV/DIVISION

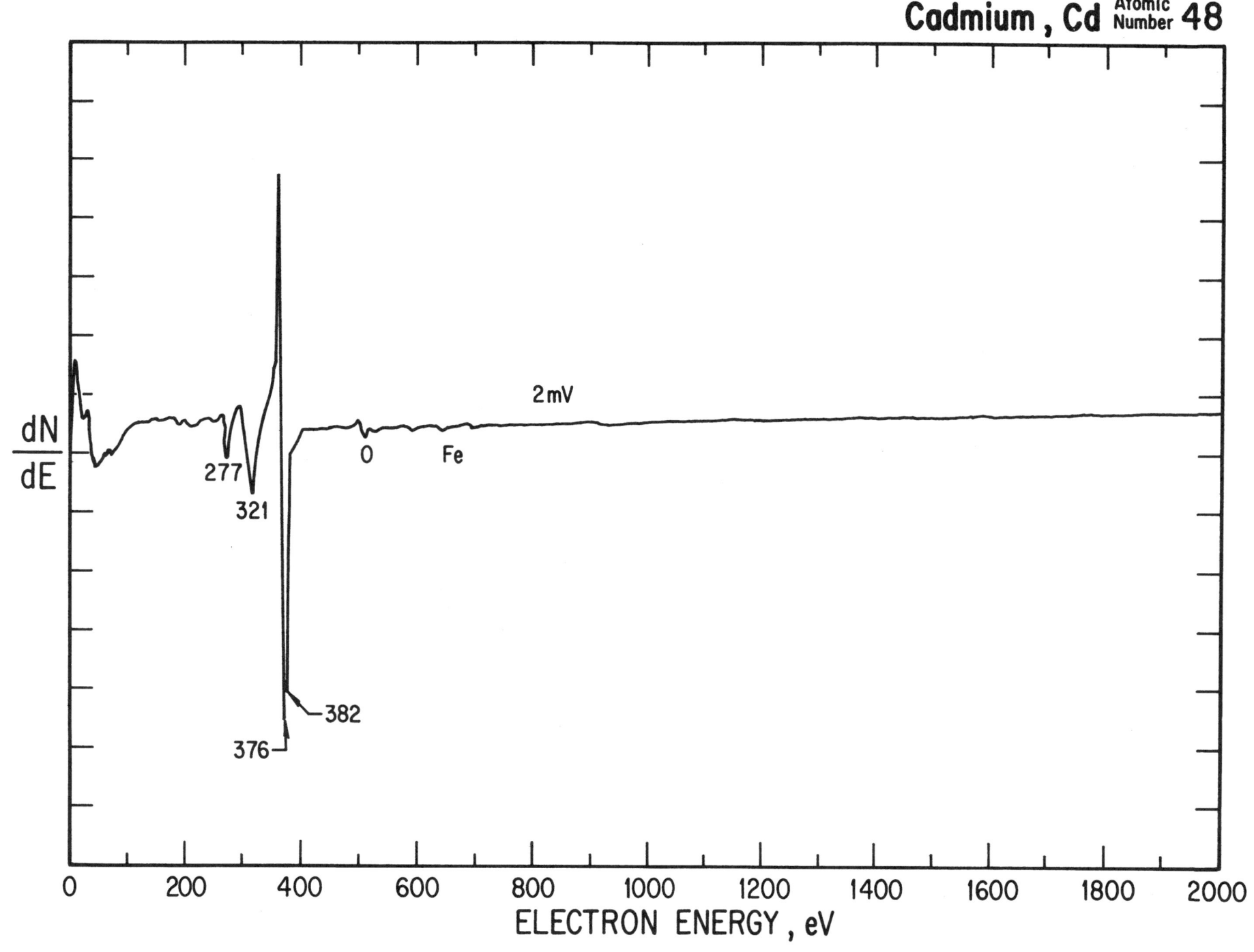

Cadmium, Cd Atomic Number 48

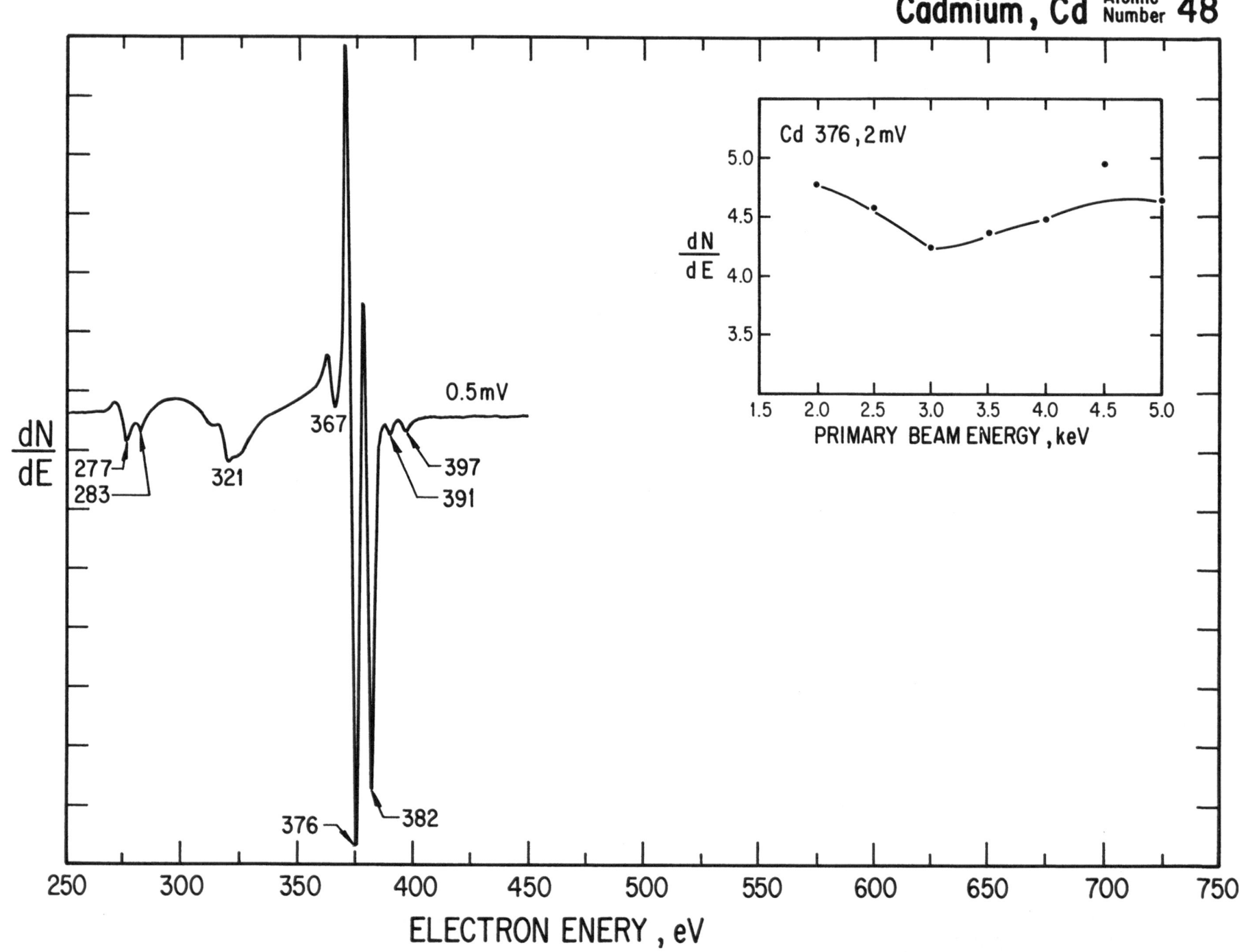

Cadmium, Cd Atomic Number **48**

Cd 376, 2 mV

$\frac{dN}{dE}$

PRIMARY BEAM ENERGY, keV

$\frac{dN}{dE}$

0.5 mV

277
283
321
367
376
382
391
397

ELECTRON ENERY, eV

Cadmium, Cd Atomic Number 48

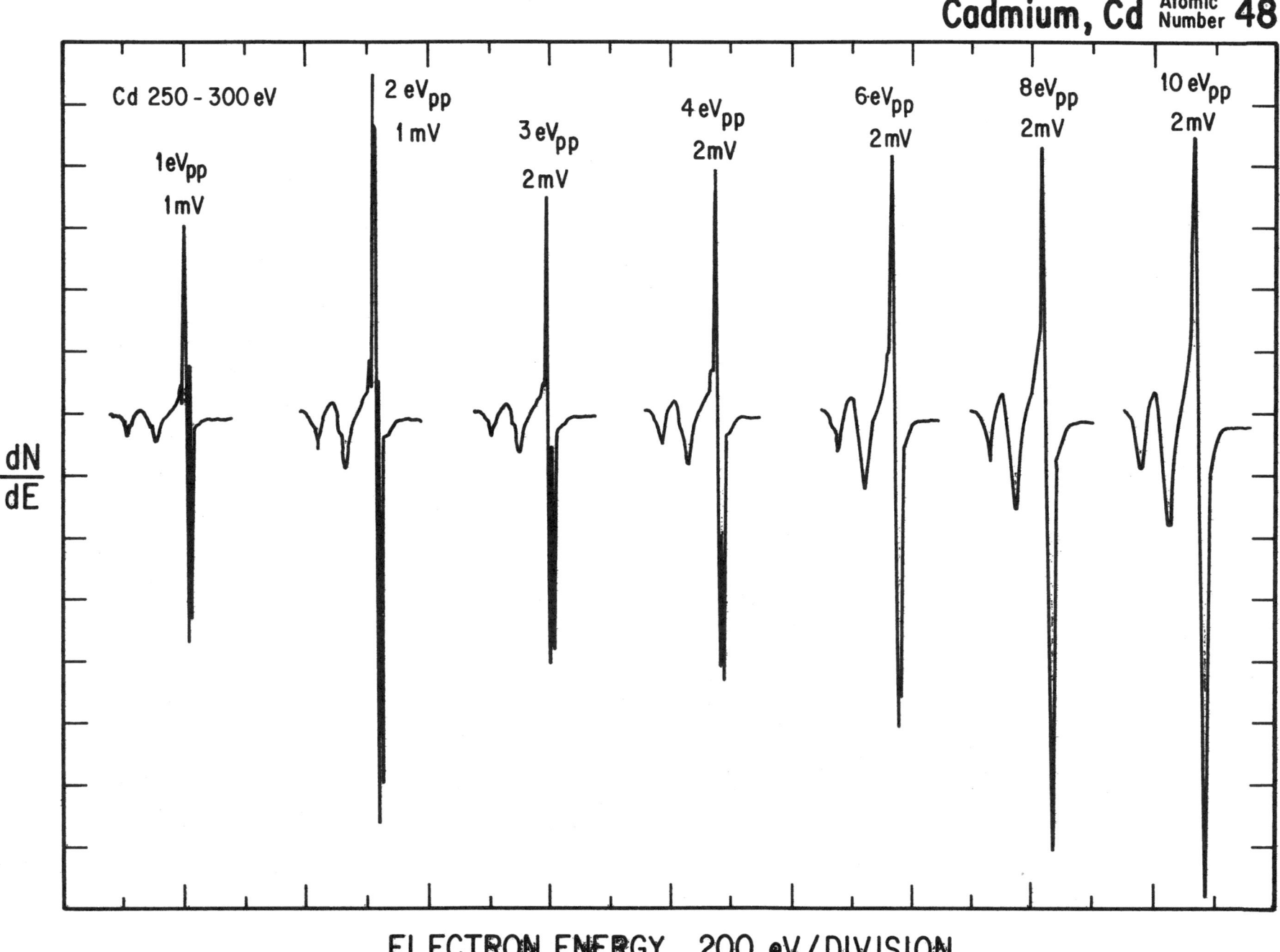

Cd 250 - 300 eV

1eV$_{pp}$
1mV

2 eV$_{pp}$
1 mV

3 eV$_{pp}$
2 mV

4 eV$_{pp}$
2mV

6 eV$_{pp}$
2 mV

8eV$_{pp}$
2mV

10 eV$_{pp}$
2mV

$\frac{dN}{dE}$

ELECTRON ENERGY, 200 eV/DIVISION

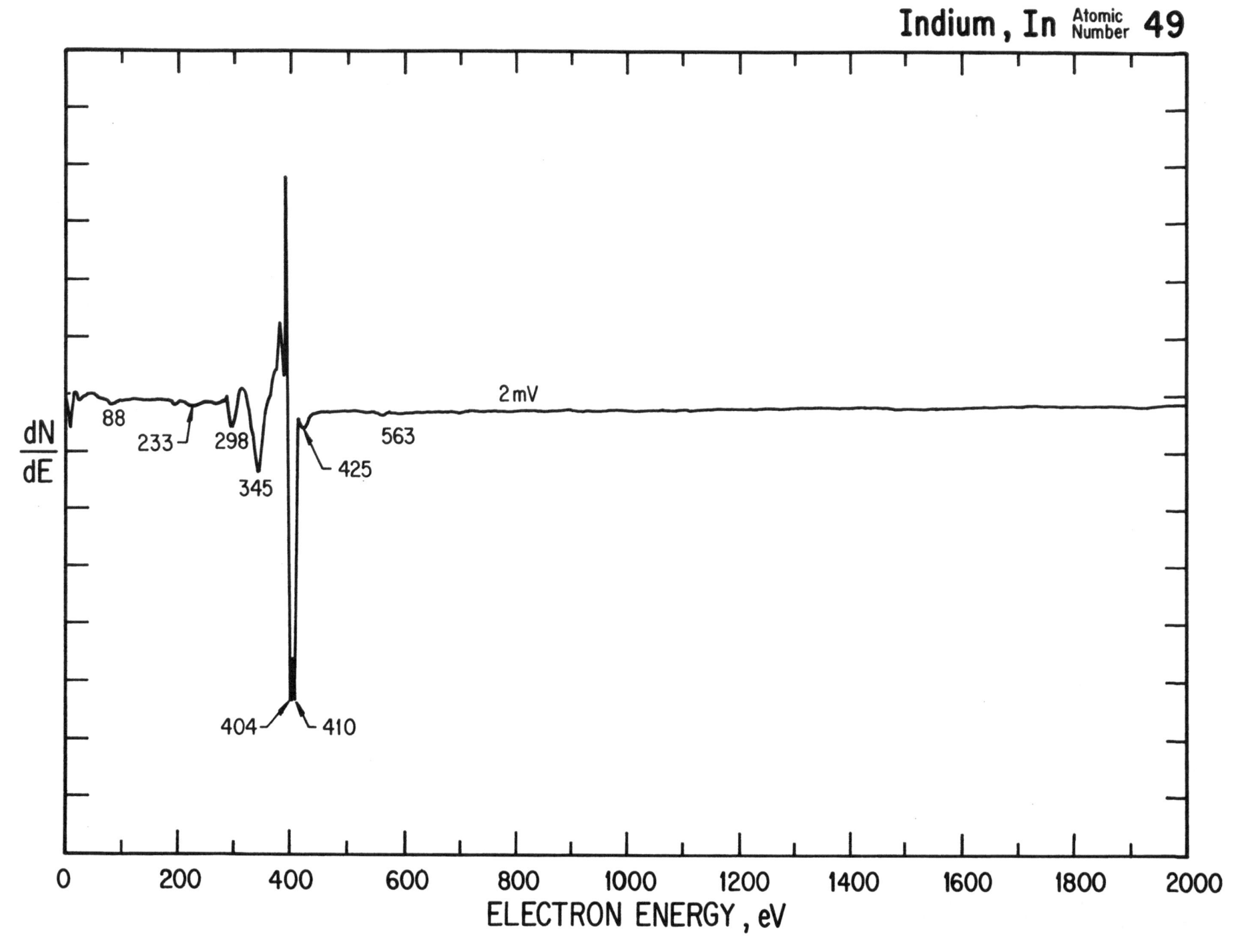

Indium, In Atomic Number **49**

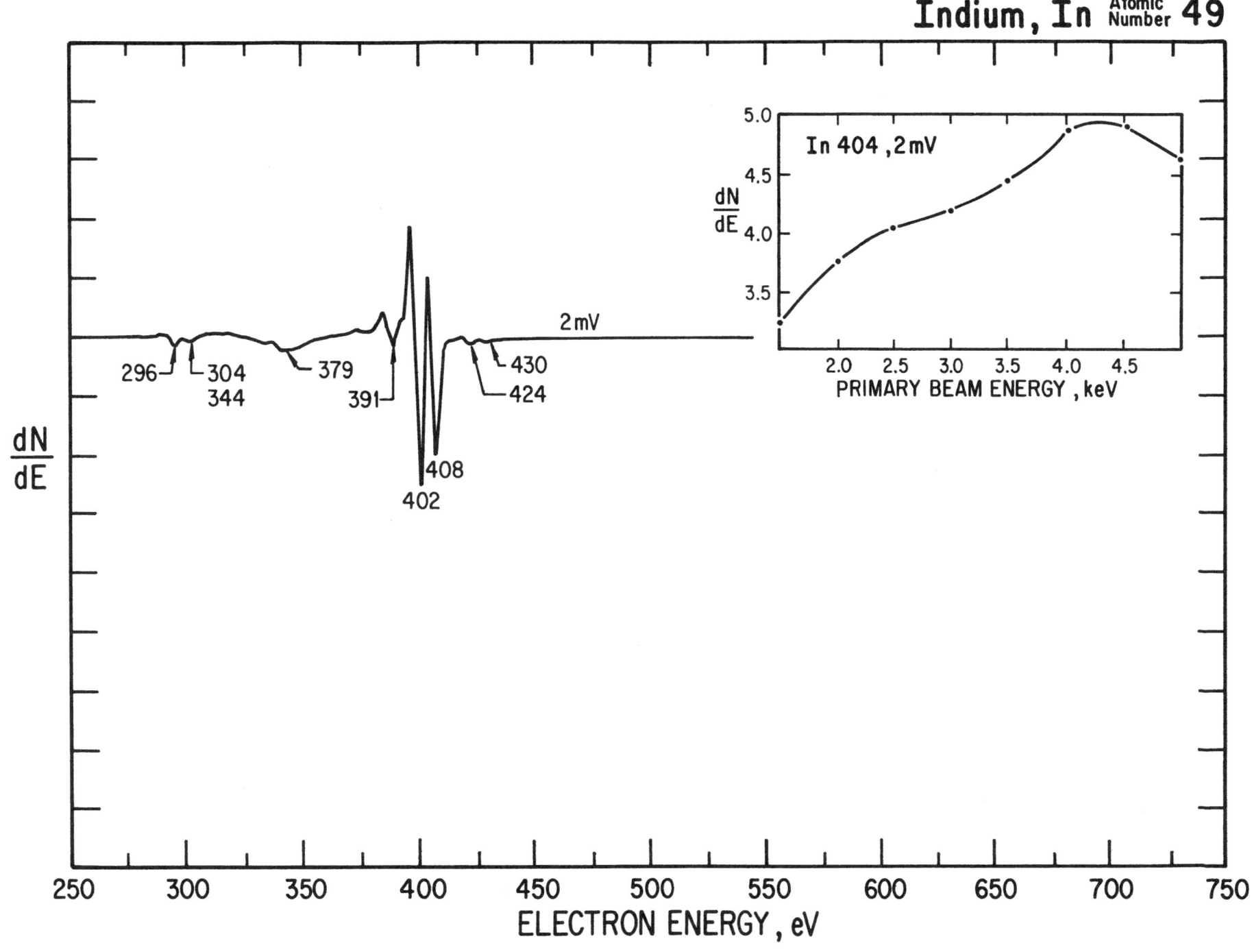

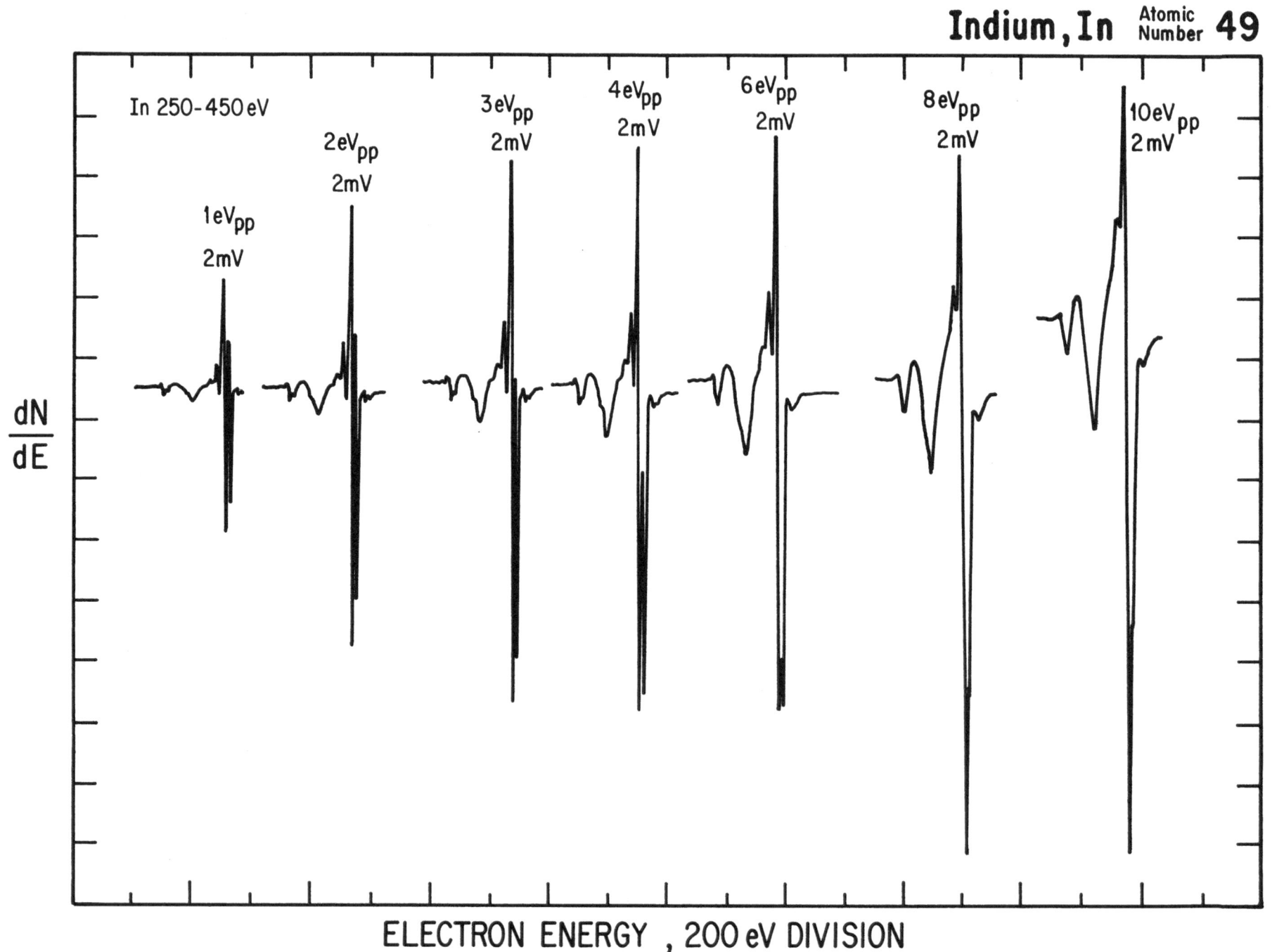

In 250-450 eV

1eV_pp
2mV

2eV_pp
2mV

3eV_pp
2mV

4eV_pp
2mV

6eV_pp
2mV

8eV_pp
2mV

10eV_pp
2mV

$\dfrac{dN}{dE}$

ELECTRON ENERGY , 200 eV DIVISION

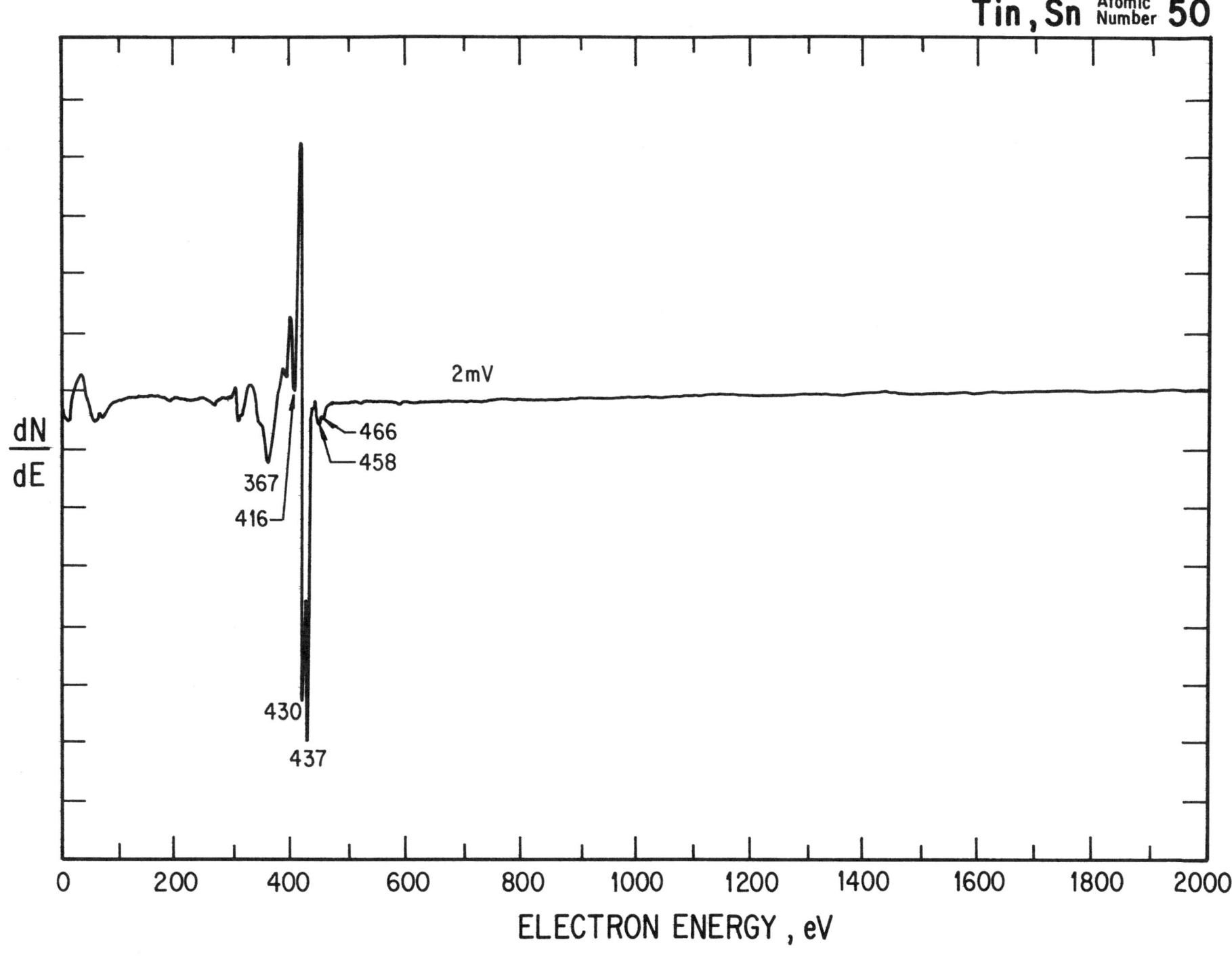

$\dfrac{dN}{dE}$

2mV

466

458

367

416

430

437

ELECTRON ENERGY , eV

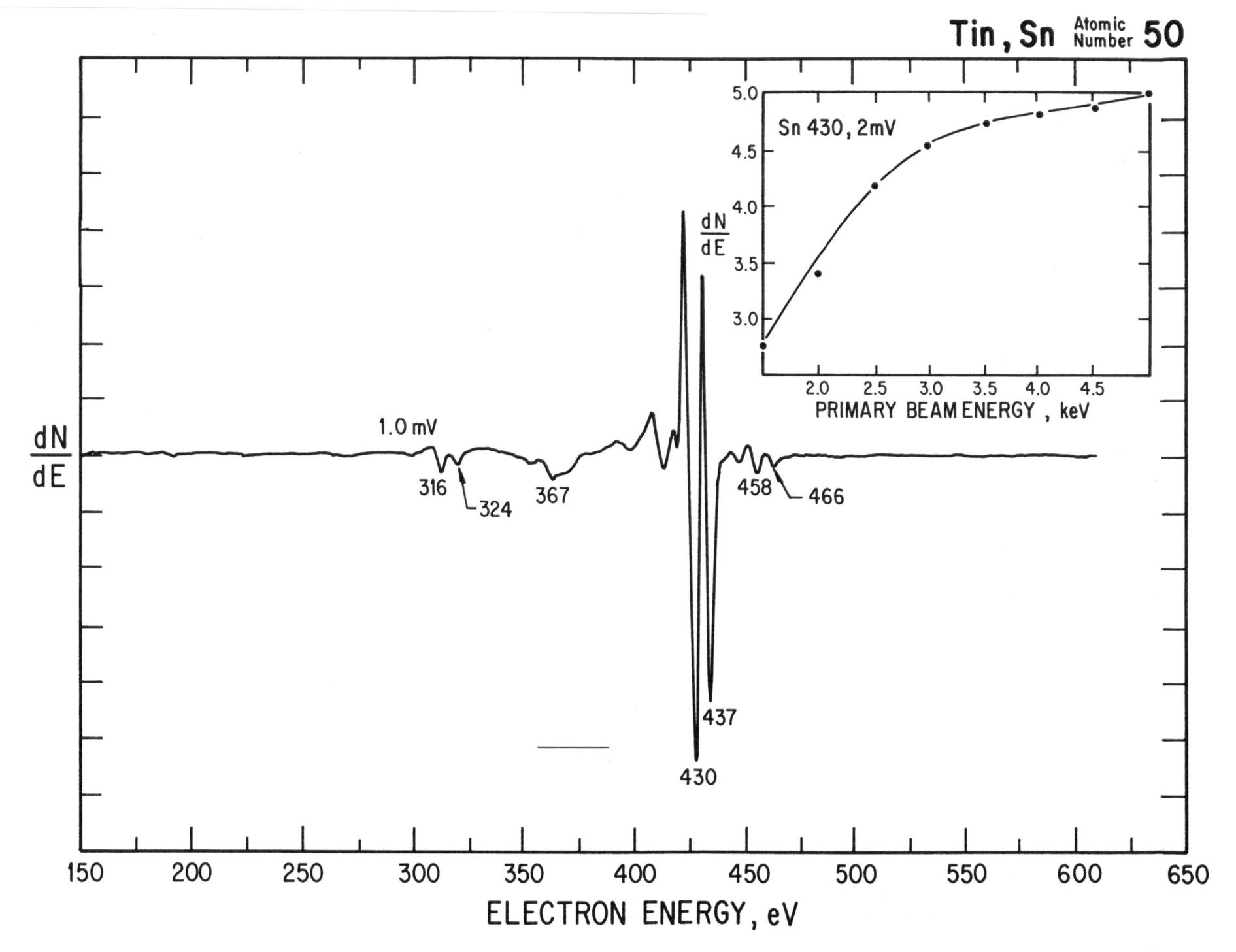

dN/dE

1.0 mV

316
324
367

437

430

458
466

Sn 430, 2mV

dN/dE

5.0

4.5

4.0

3.5

3.0

2.0   2.5   3.0   3.5   4.0   4.5

PRIMARY BEAM ENERGY , keV

150   200   250   300   350   400   450   500   550   600   650

ELECTRON ENERGY, eV

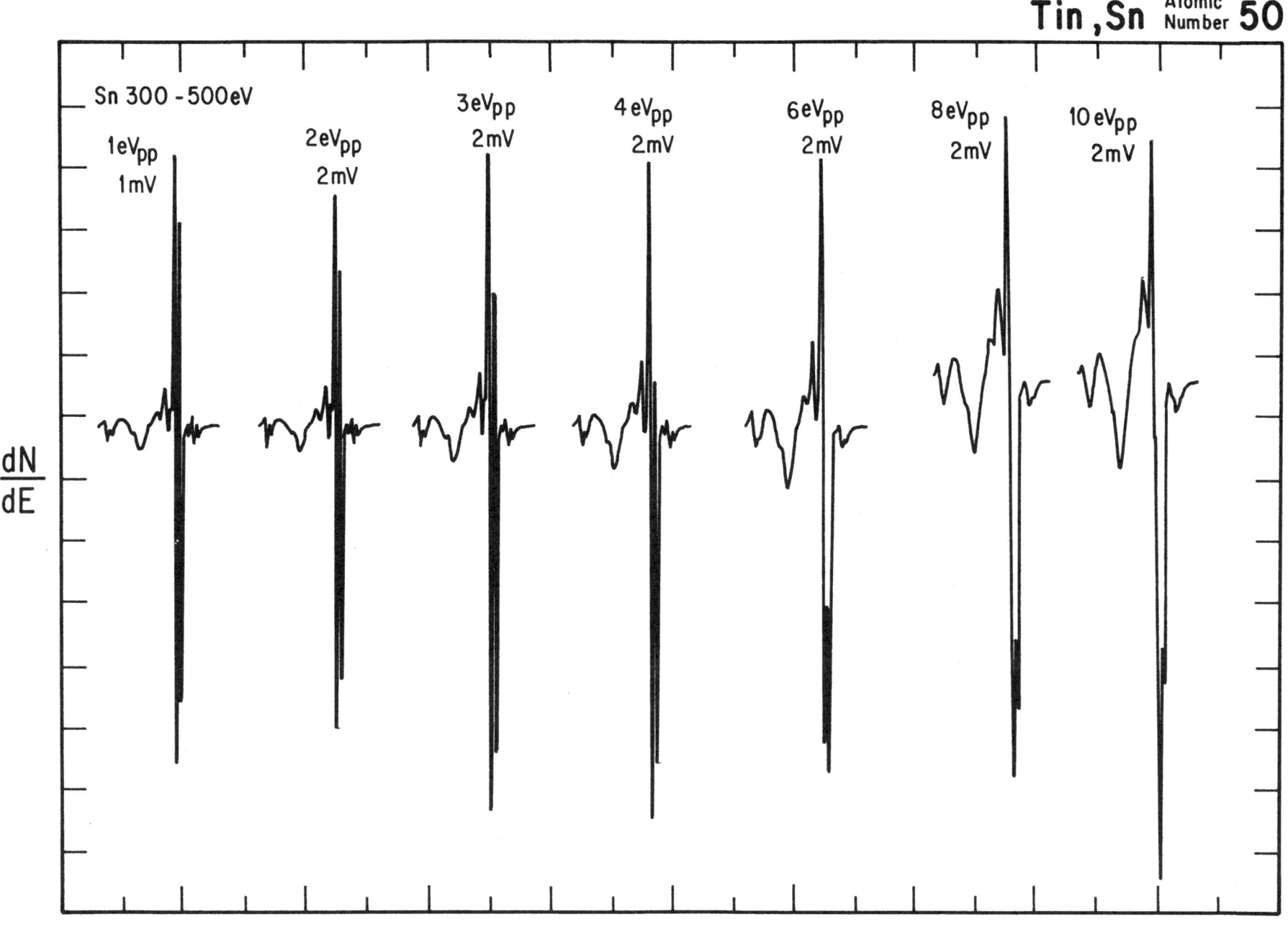

Sn 300 - 500 eV

1 eV$_{pp}$
1 mV

2 eV$_{pp}$
2 mV

3 eV$_{pp}$
2 mV

4 eV$_{pp}$
2 mV

6 eV$_{pp}$
2 mV

8 eV$_{pp}$
2 mV

10 eV$_{pp}$
2 mV

$\dfrac{dN}{dE}$

ELECTRON ENERGY, 200 eV/DIVISION

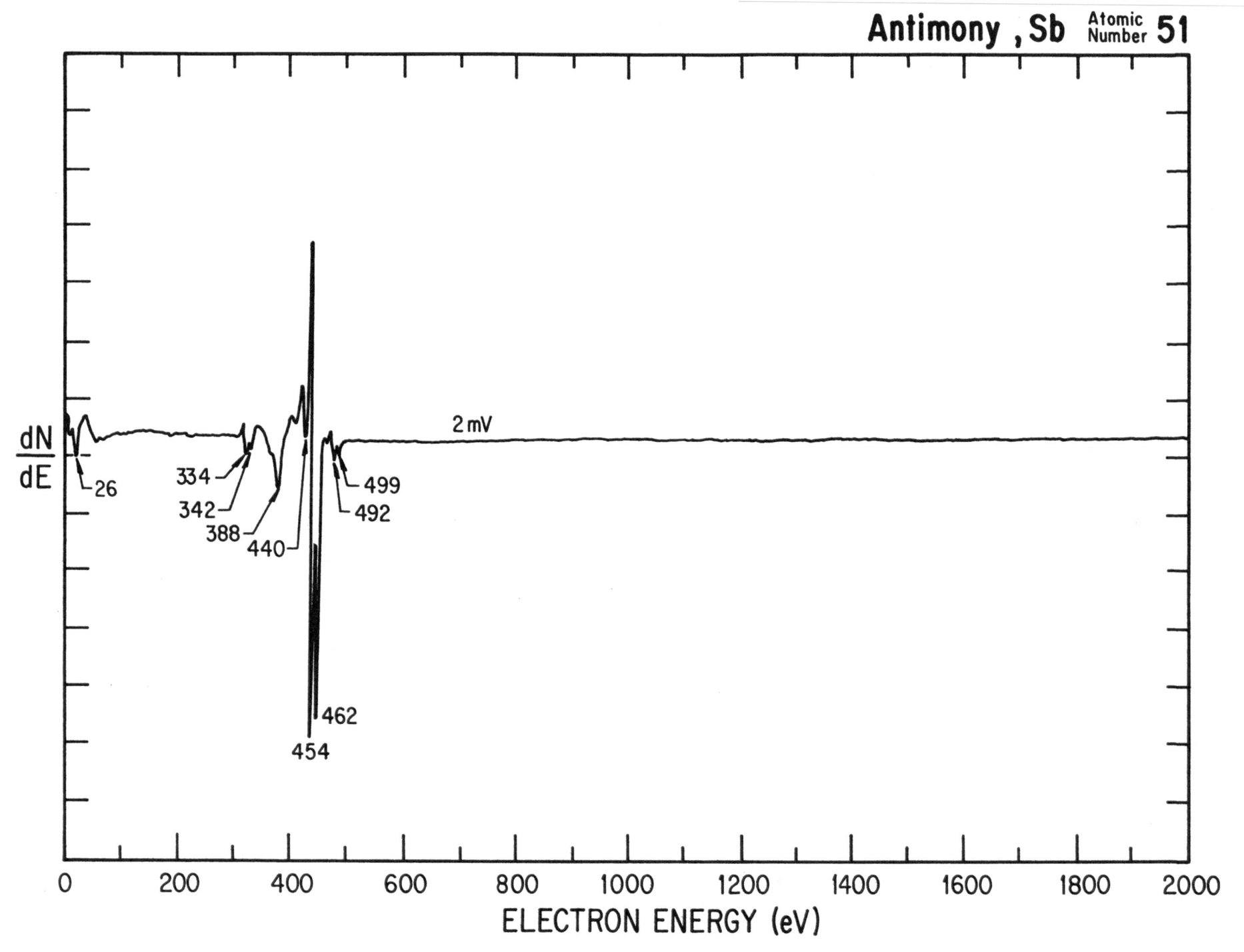

Antimony , Sb Atomic Number **51**

$\dfrac{dN}{dE}$

26

334

342

388

440

2 mV

499

492

454 462

ELECTRON ENERGY (eV)

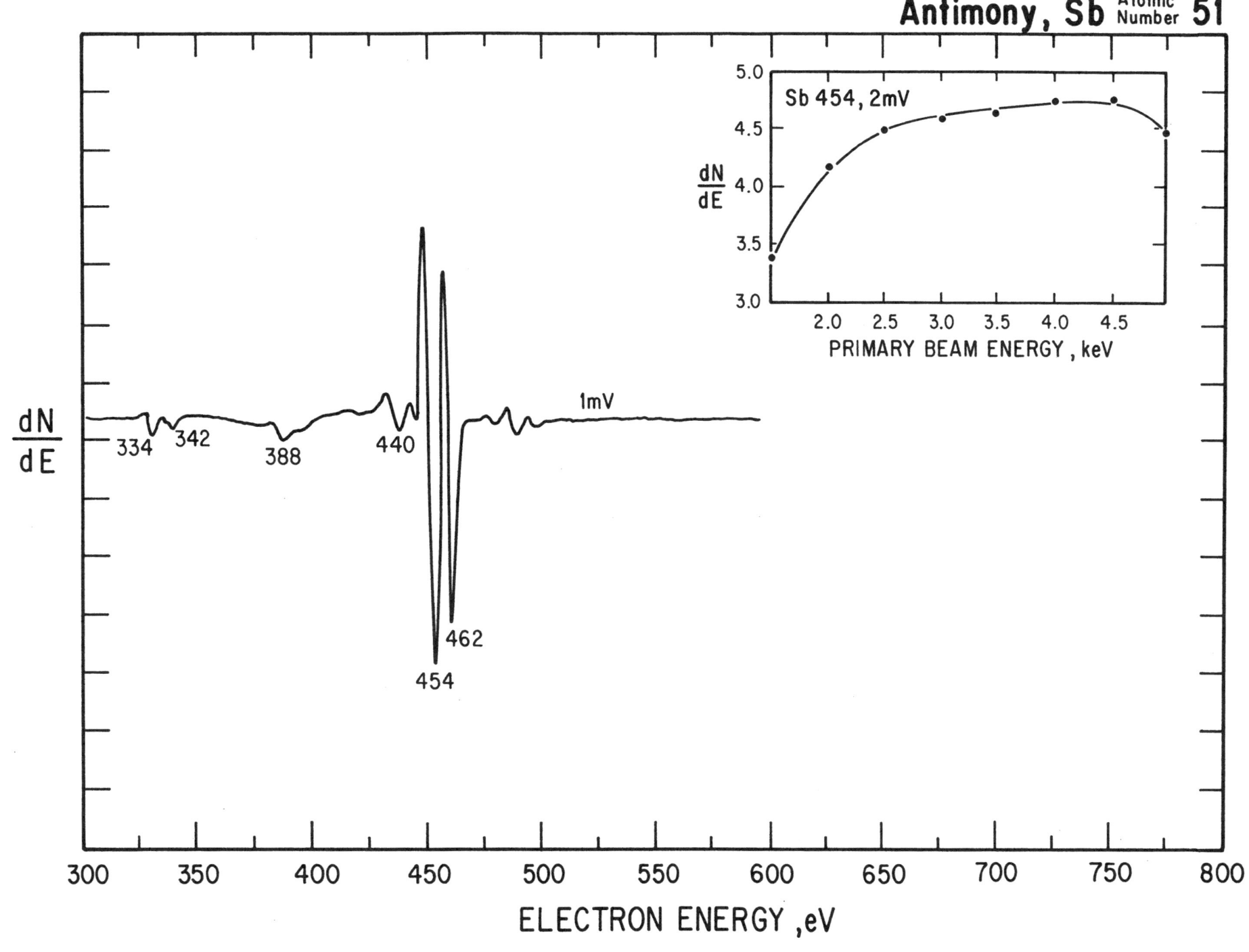

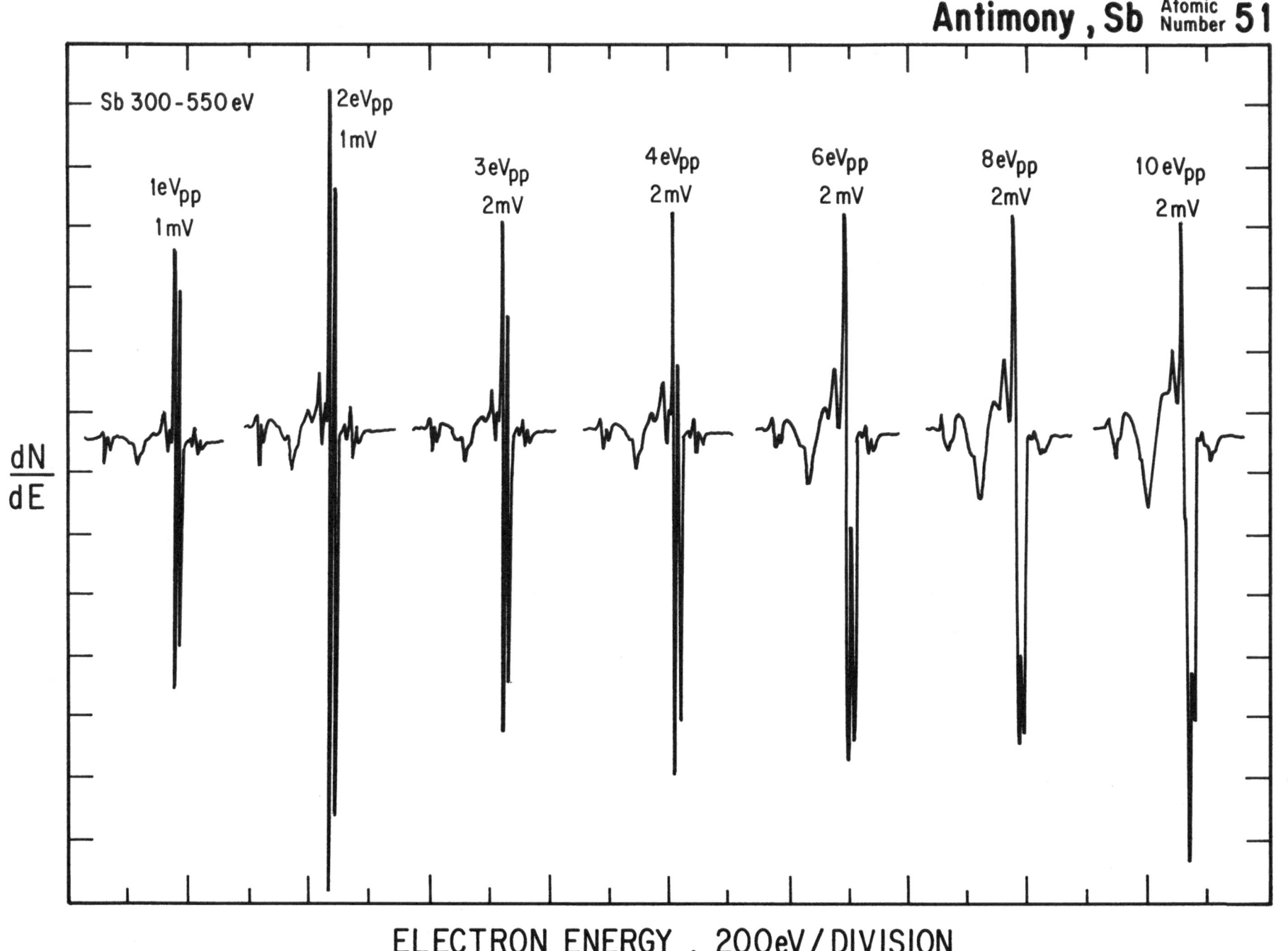

Antimony , Sb Atomic Number **51**

Sb 300-550 eV

1eV$_{pp}$
1mV

2eV$_{pp}$
1mV

3eV$_{pp}$
2mV

4eV$_{pp}$
2mV

6eV$_{pp}$
2mV

8eV$_{pp}$
2mV

10eV$_{pp}$
2mV

$\dfrac{dN}{dE}$

ELECTRON ENERGY , 200eV / DIVISION

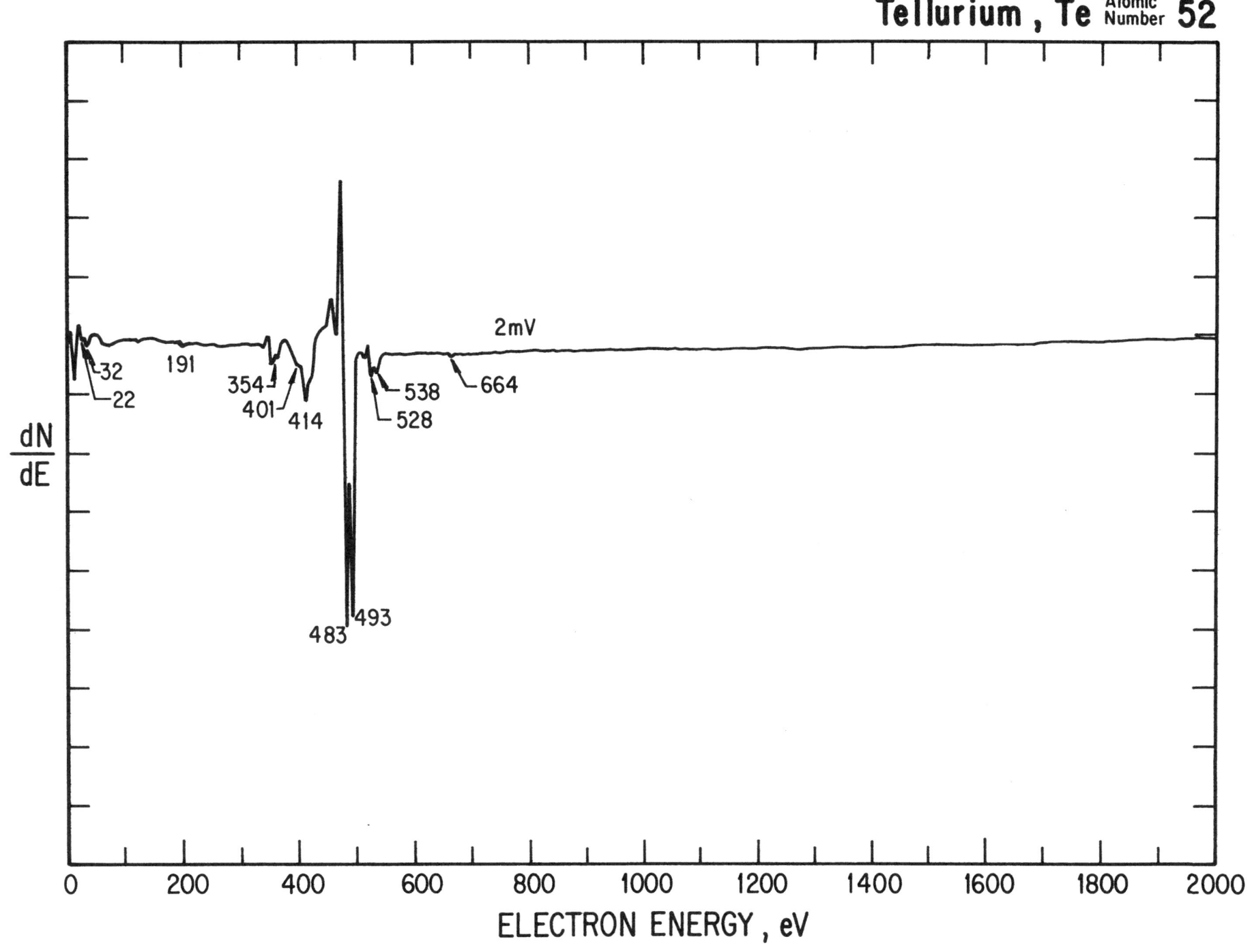

2mV

32
22
191
354
401
414
483
493
538
528
664

$\frac{dN}{dE}$

ELECTRON ENERGY , eV

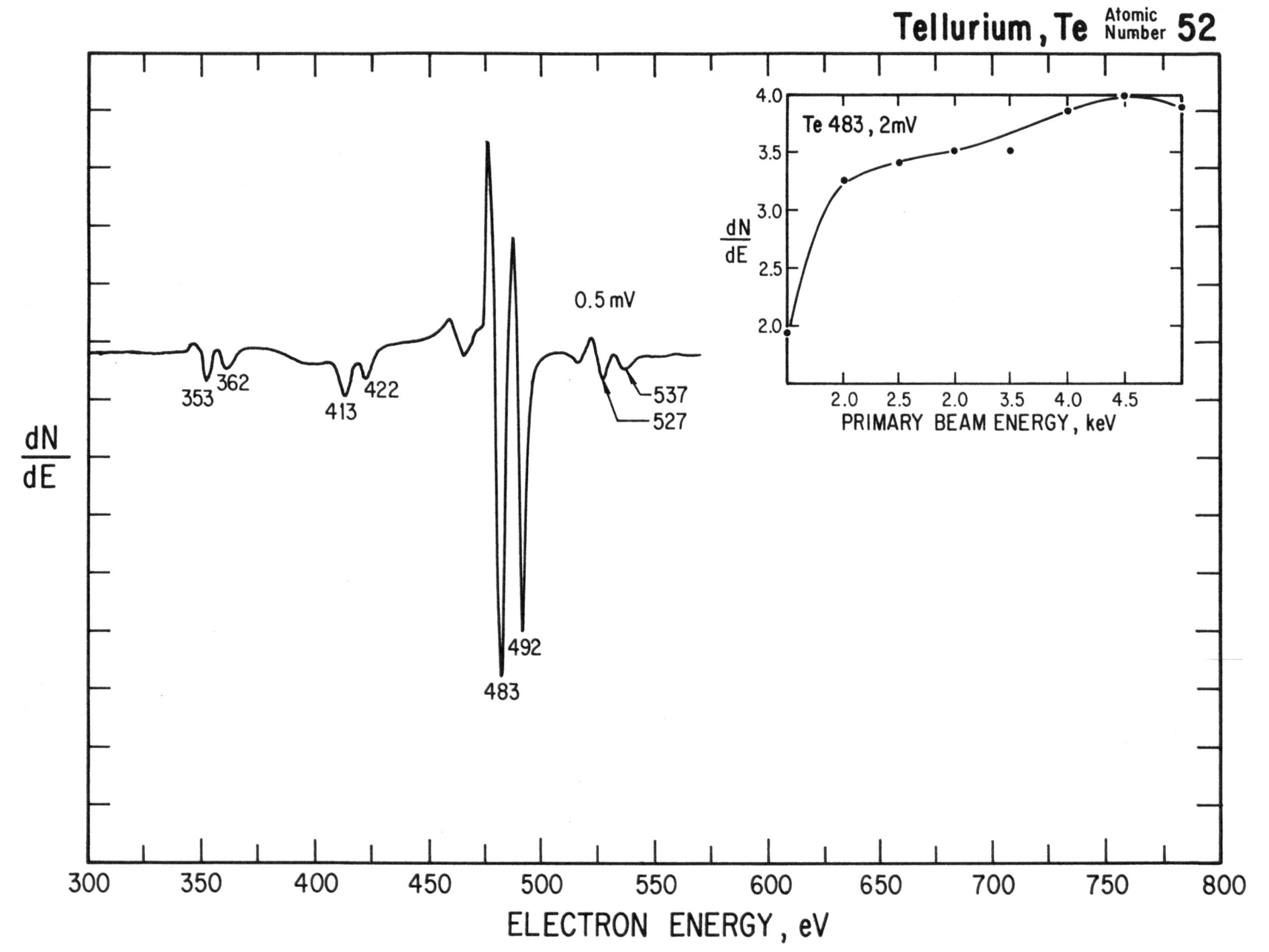

Tellurium, Te Atomic Number 52

$\dfrac{dN}{dE}$

ELECTRON ENERGY, eV

0.5 mV

353 362 413 422 483 492 527 537

Te 483, 2mV

$\dfrac{dN}{dE}$

PRIMARY BEAM ENERGY, keV

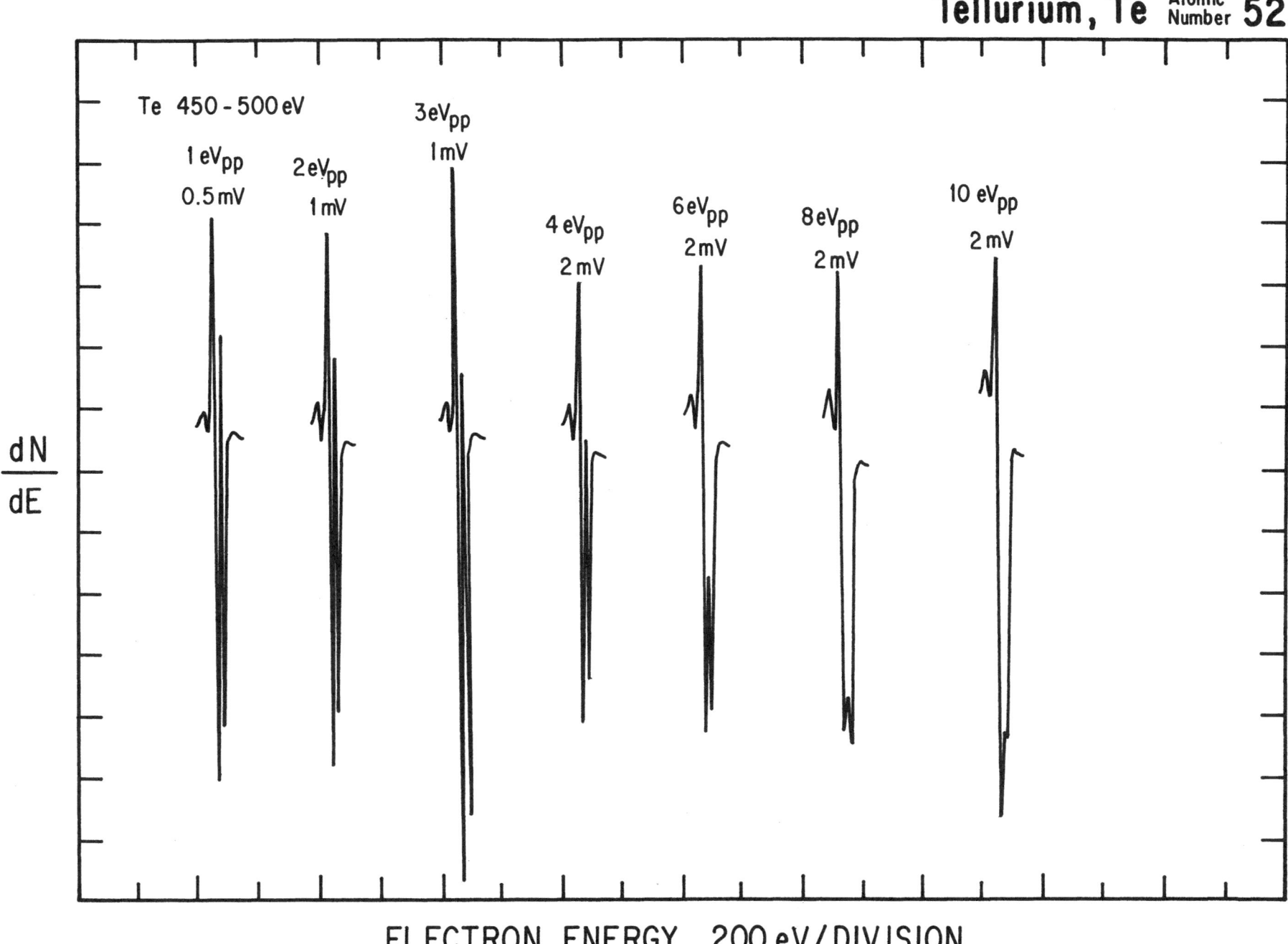

Te 450 - 500 eV

1 eVpp
0.5 mV

2 eVpp
1 mV

3 eVpp
1 mV

4 eVpp
2 mV

6 eVpp
2 mV

8 eVpp
2 mV

10 eVpp
2 mV

$\dfrac{dN}{dE}$

ELECTRON ENERGY , 200 eV/DIVISION

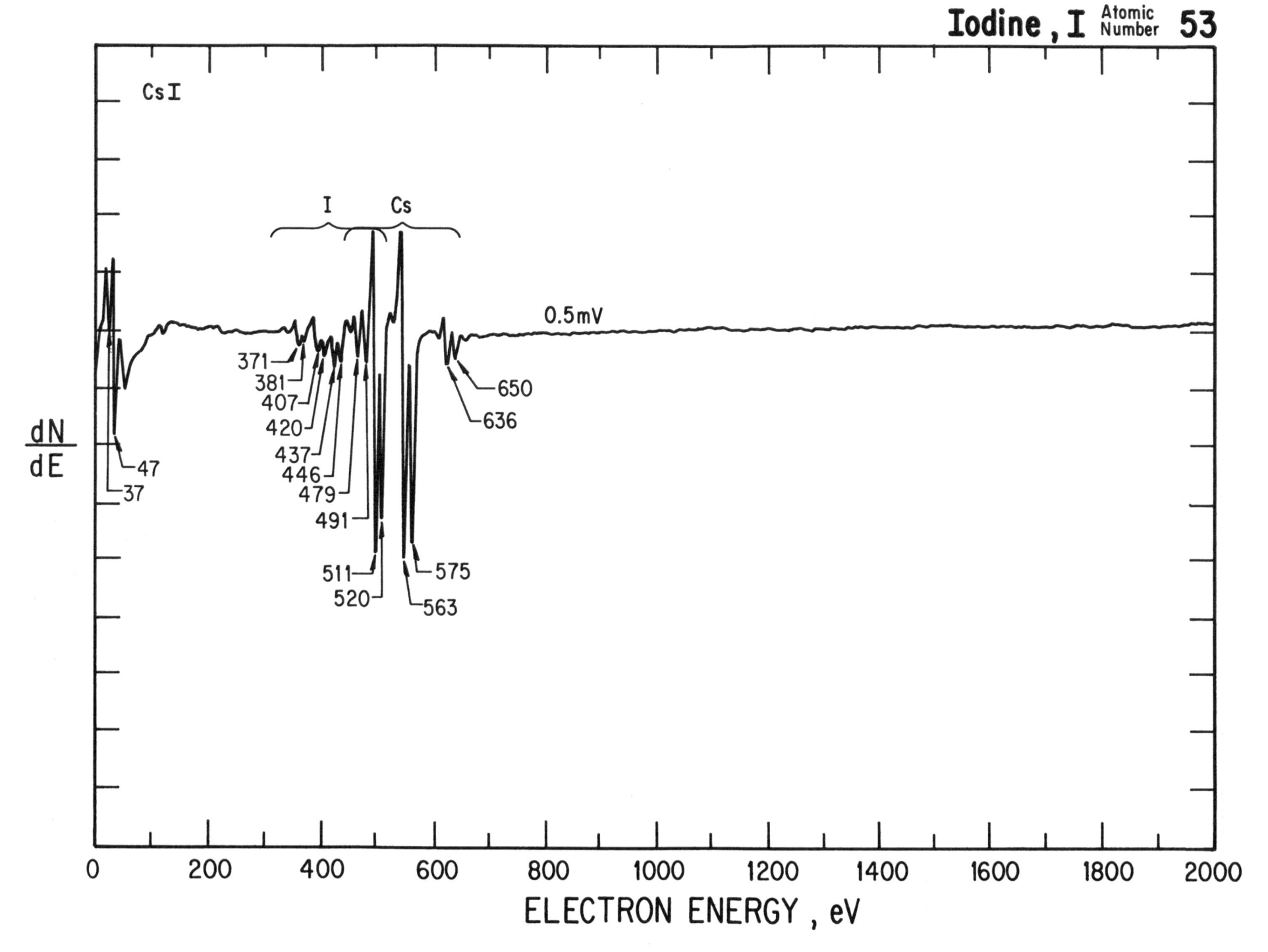

Iodine, I Atomic Number 53

CsI

I   Cs

0.5mV

$\dfrac{dN}{dE}$

47
37

371
381
407
420
437
446
479
491
511
520

650
636

575
563

ELECTRON ENERGY, eV

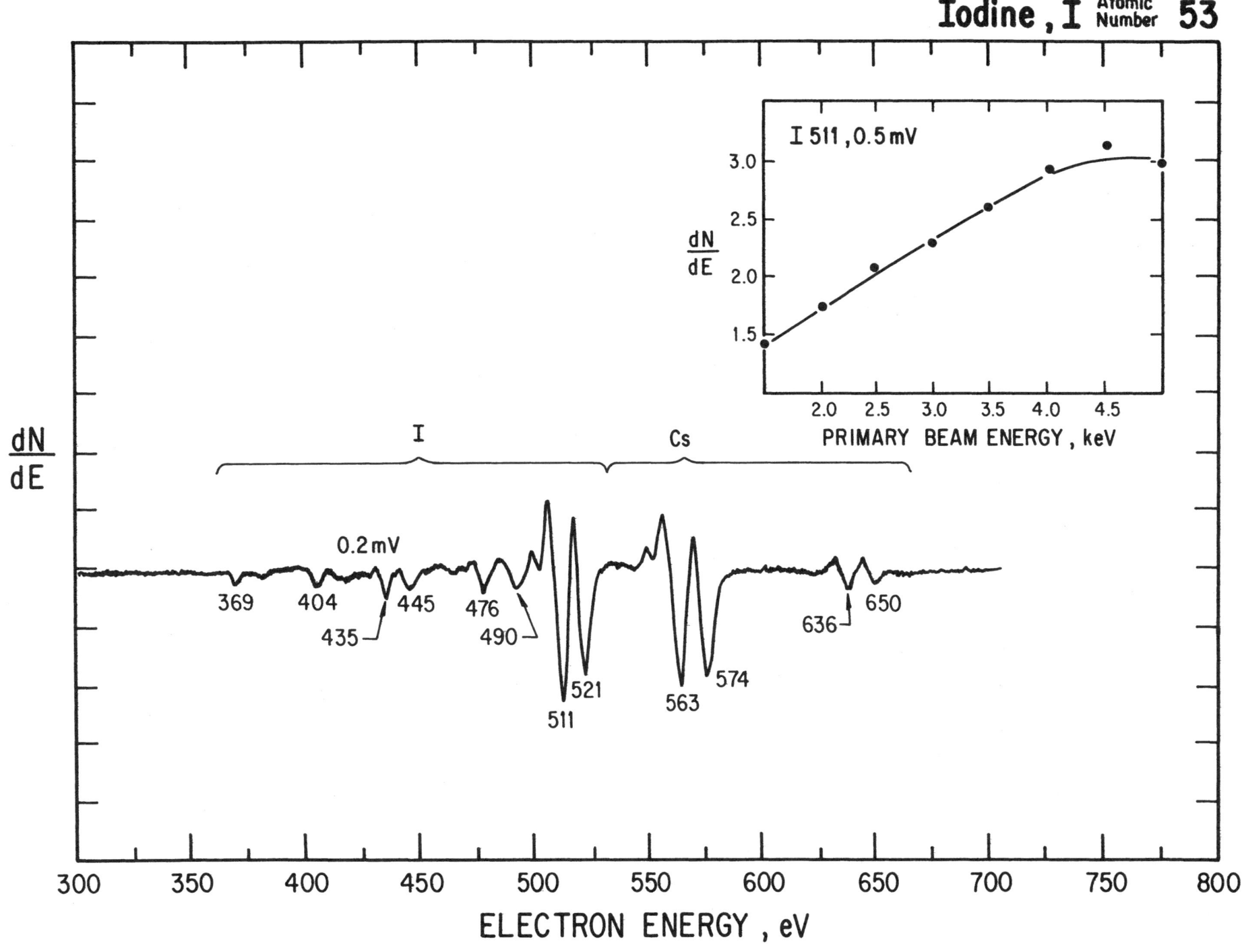

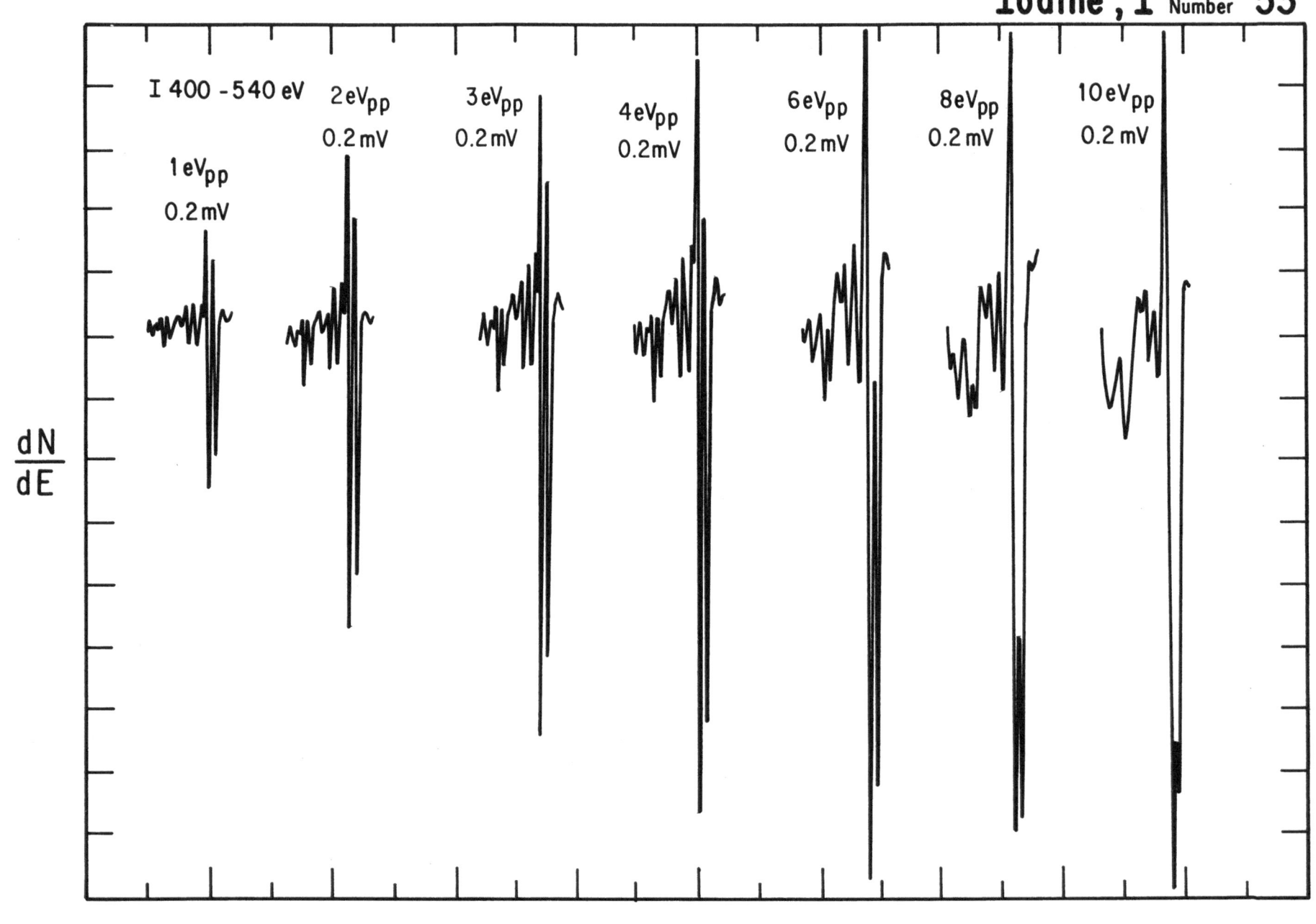

Iodine, I Atomic Number 53

$\frac{dN}{dE}$

I 400 - 540 eV

1 eV_pp
0.2 mV

2 eV_pp
0.2 mV

3 eV_pp
0.2 mV

4 eV_pp
0.2 mV

6 eV_pp
0.2 mV

8 eV_pp
0.2 mV

10 eV_pp
0.2 mV

ELECTRON ENERGY , 200 eV / DIVISION

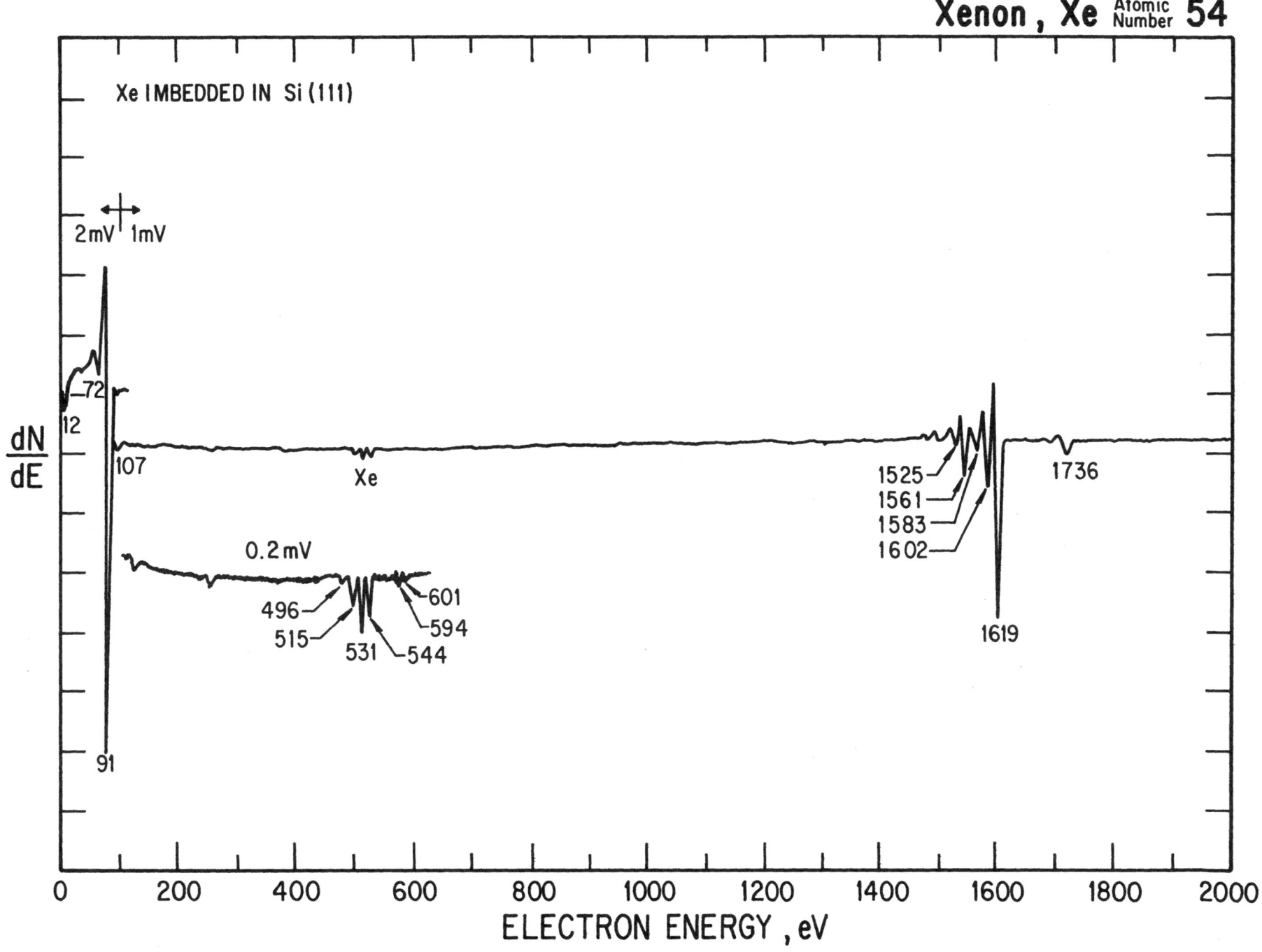

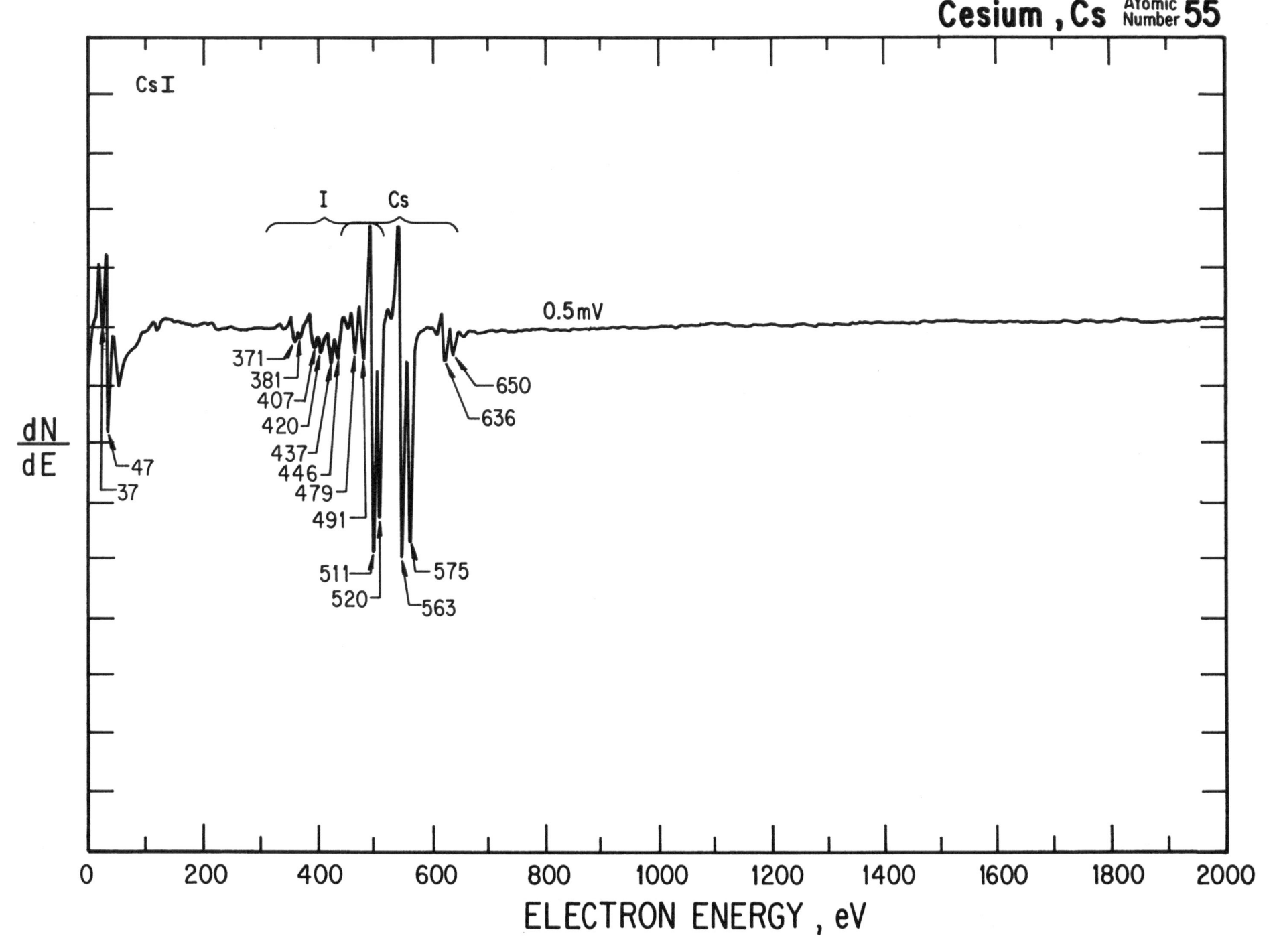

Cesium , Cs Atomic Number 55

CsI

I   Cs

0.5mV

$\dfrac{dN}{dE}$

47
37

371
381
407
420
437
446
479
491

511
520

575
563

650
636

ELECTRON ENERGY , eV

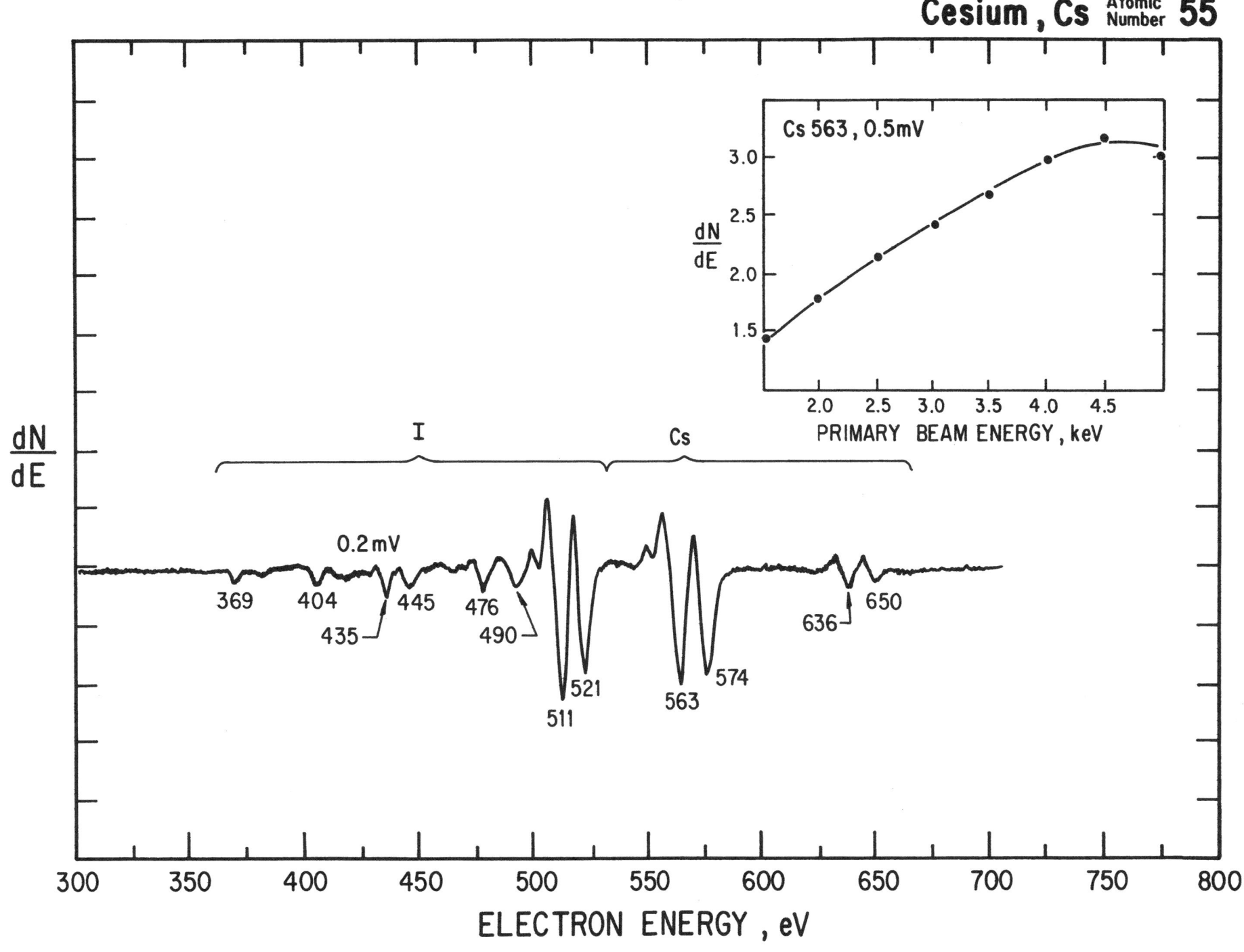

Cesium, Cs Atomic Number **55**

Cesium, Cs Atomic Number 55

Cs 540 - 700 eV

1 eVpp 0.2 mV

2 eVpp 0.2 mV

3 eVpp 0.2 mV

4 eVpp 0.2 mV

6 eVpp 0.5 mV

8 eVpp 0.5 mV

10 eVpp 0.5 mV

$\frac{dN}{dE}$

ELECTRON ENERGY, 200 eV/DIVISION

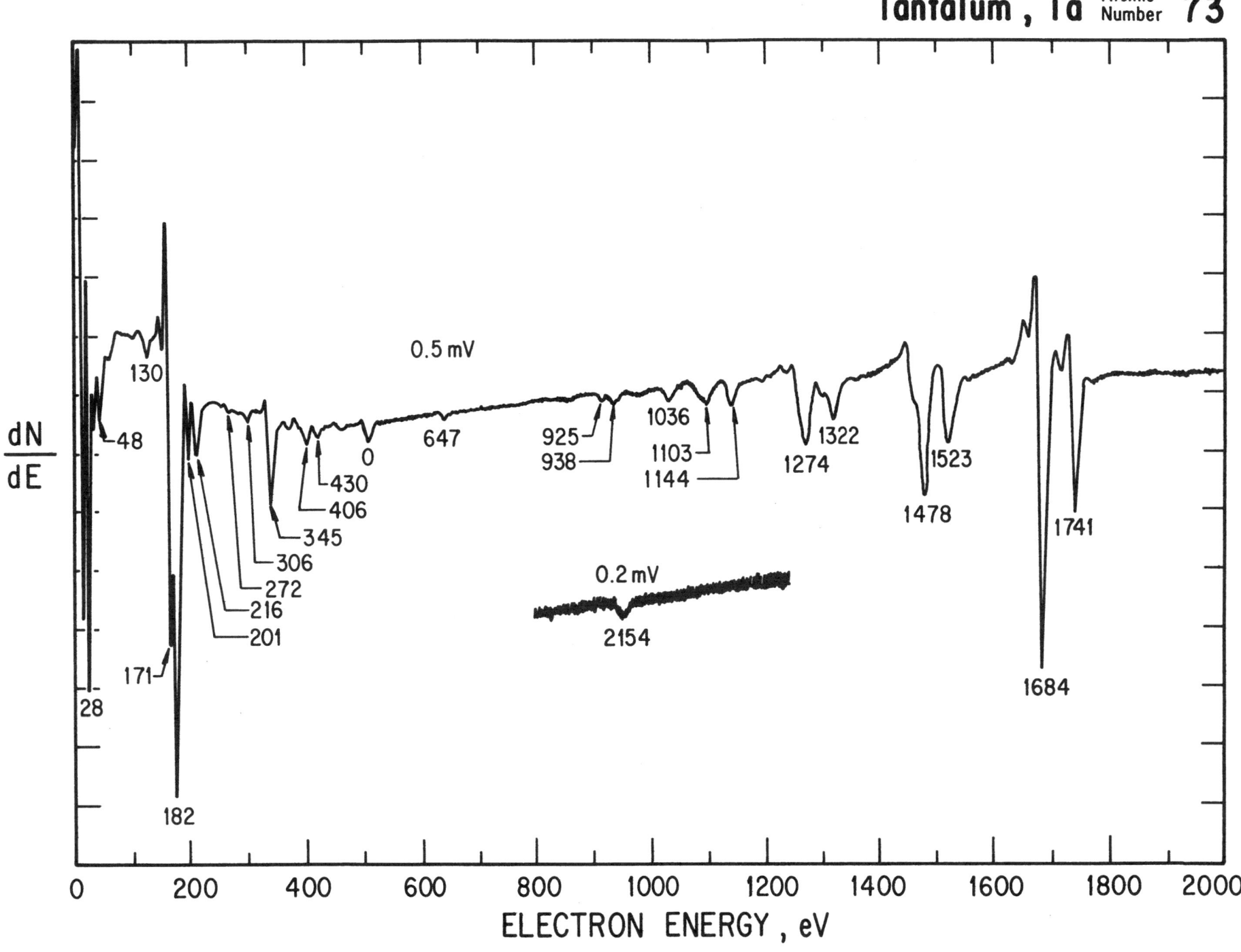

ELECTRON ENERGY, eV

$\dfrac{dN}{dE}$

0.5 mV

0.2 mV

28

130

48

171

182

430
406
345
306
272
216
201

0

647

925
938

1036
1103
1144

1274

1322

1478

1523

1684

1741

2154

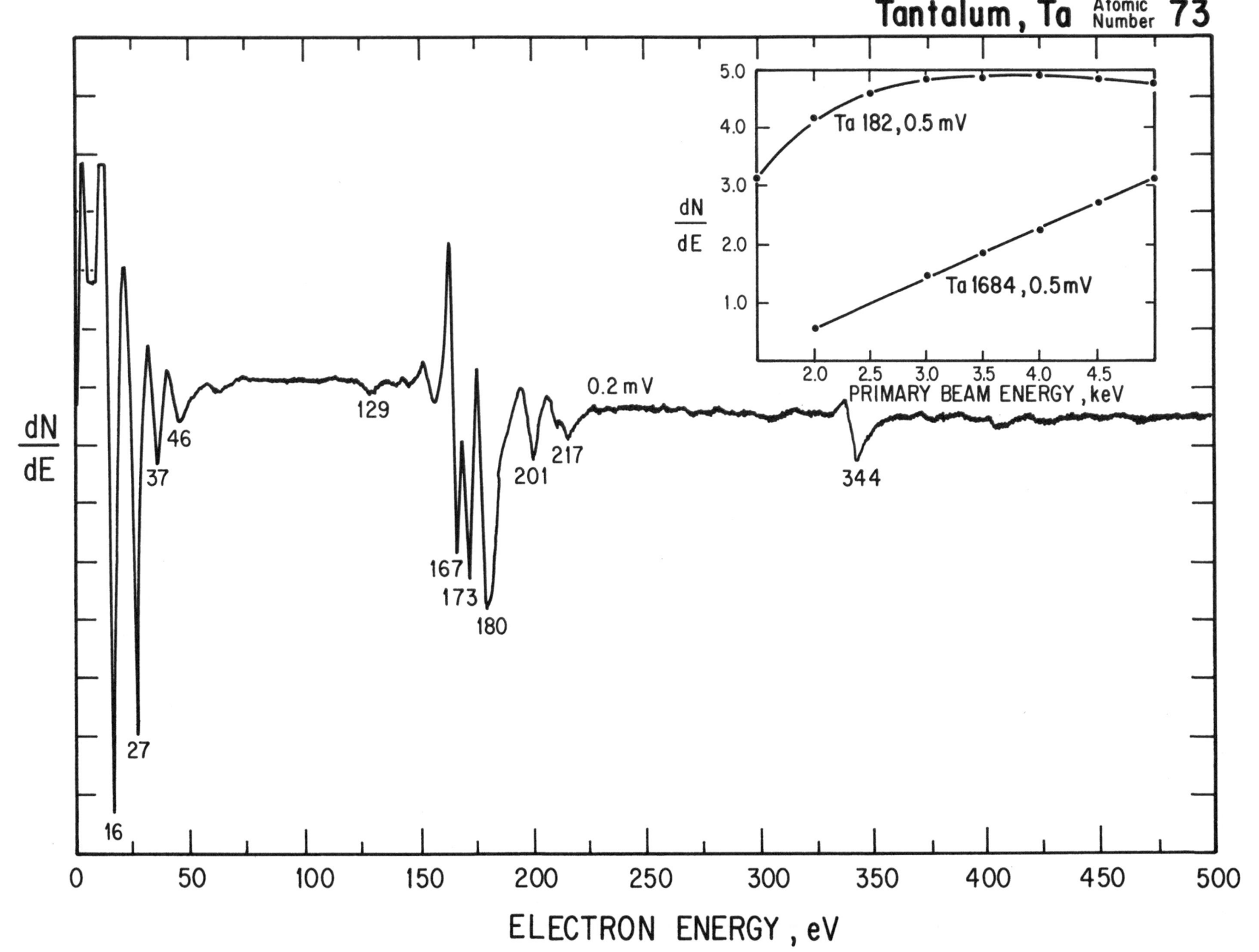

$\dfrac{dN}{dE}$

16
27
37
46
129
167
173
180
201
217
0.2 mV
344

ELECTRON ENERGY, eV

$\dfrac{dN}{dE}$

Ta 182, 0.5 mV
Ta 1684, 0.5 mV

PRIMARY BEAM ENERGY, keV

5.0
4.0
3.0
2.0
1.0

2.0   2.5   3.0   3.5   4.0   4.5

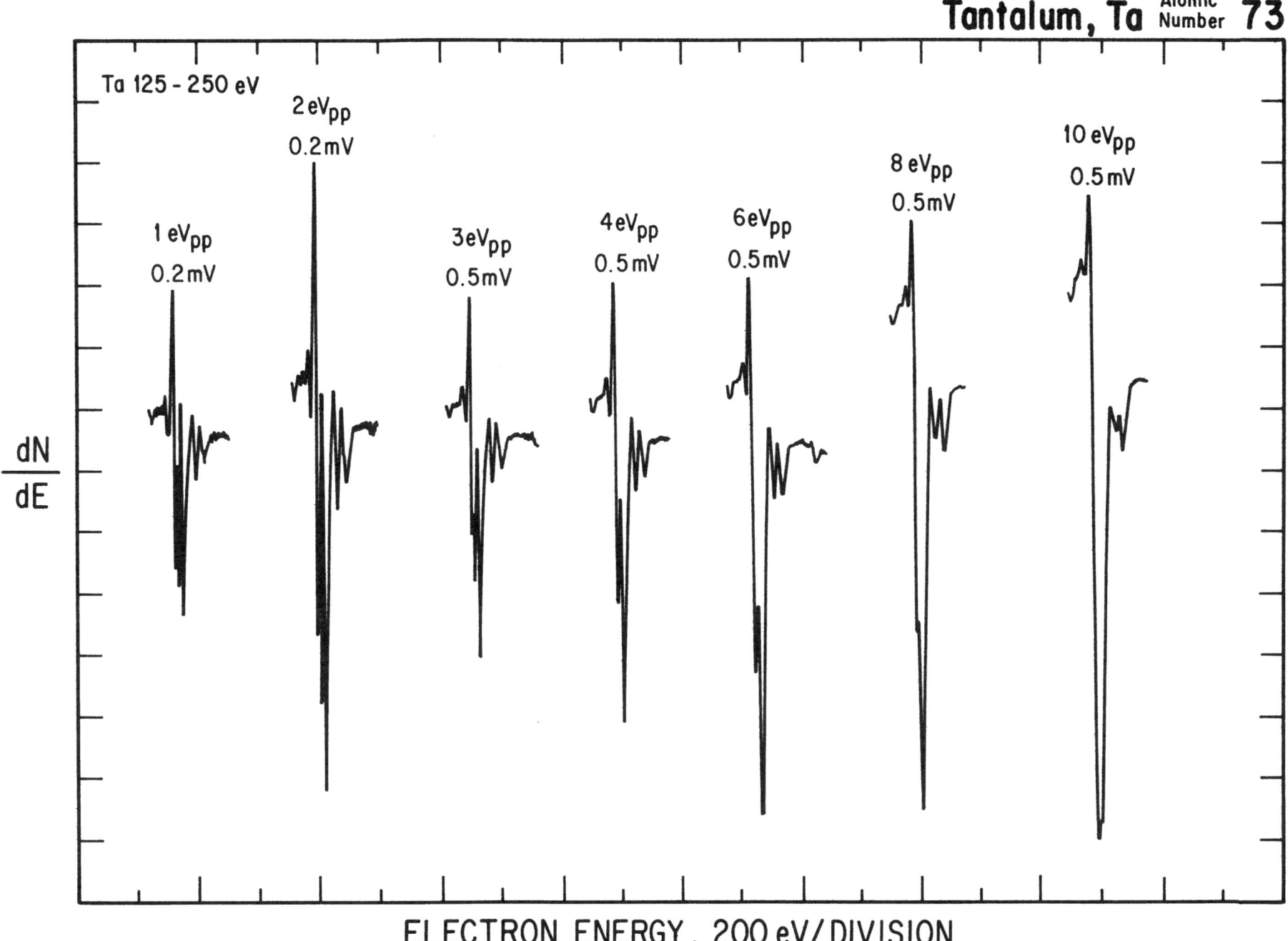

Tantalum, Ta   Atomic Number **73**

Ta 125 - 250 eV

1 eV_pp
0.2 mV

2 eV_pp
0.2 mV

3 eV_pp
0.5 mV

4 eV_pp
0.5 mV

6 eV_pp
0.5 mV

8 eV_pp
0.5 mV

10 eV_pp
0.5 mV

$\dfrac{dN}{dE}$

ELECTRON ENERGY, 200 eV/DIVISION

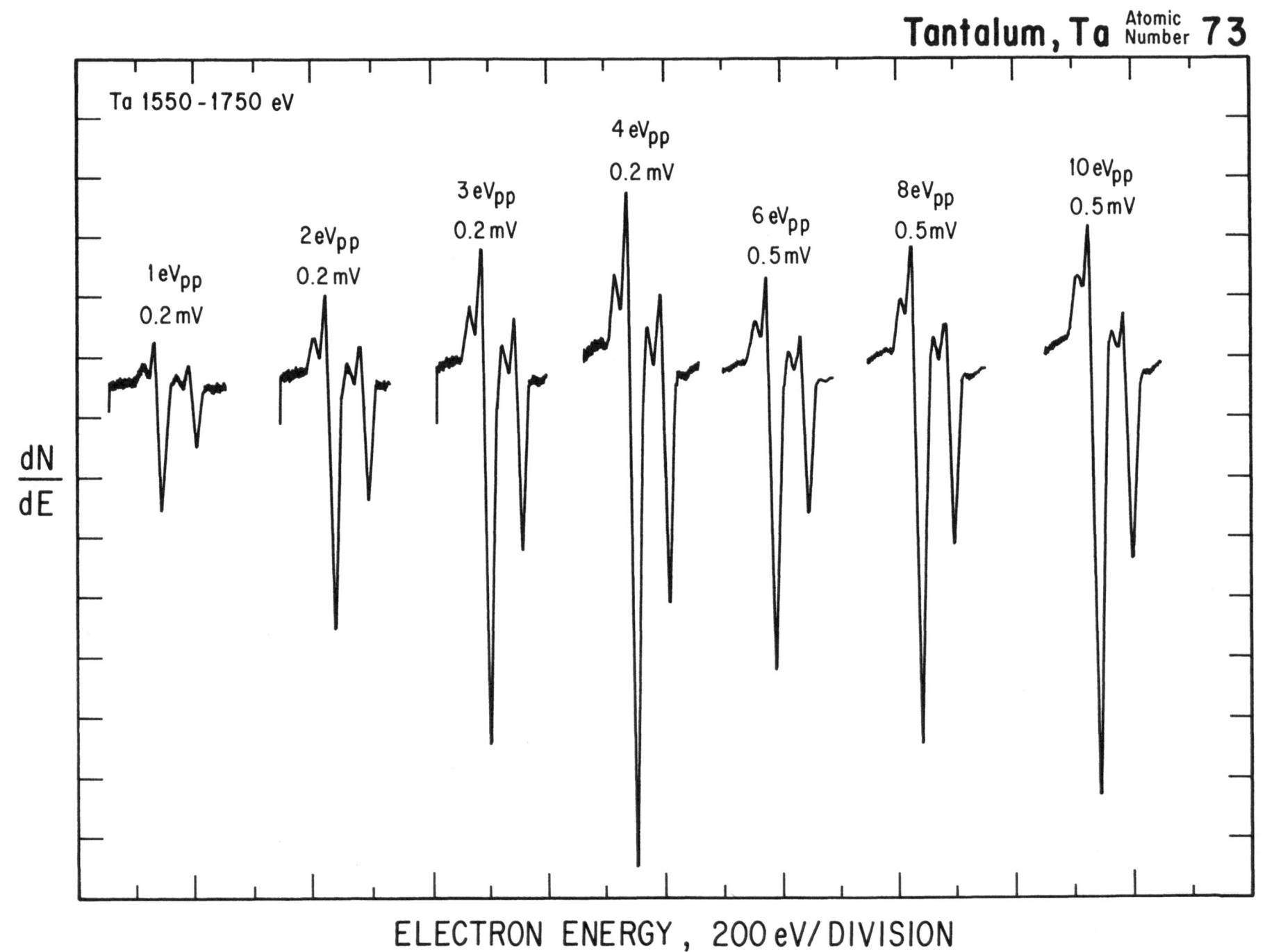

Tantalum, Ta Atomic Number 73

Ta 1550 - 1750 eV

1 eV$_{pp}$
0.2 mV

2 eV$_{pp}$
0.2 mV

3 eV$_{pp}$
0.2 mV

4 eV$_{pp}$
0.2 mV

6 eV$_{pp}$
0.5 mV

8 eV$_{pp}$
0.5 mV

10 eV$_{pp}$
0.5 mV

$\dfrac{dN}{dE}$

ELECTRON ENERGY, 200 eV/DIVISION

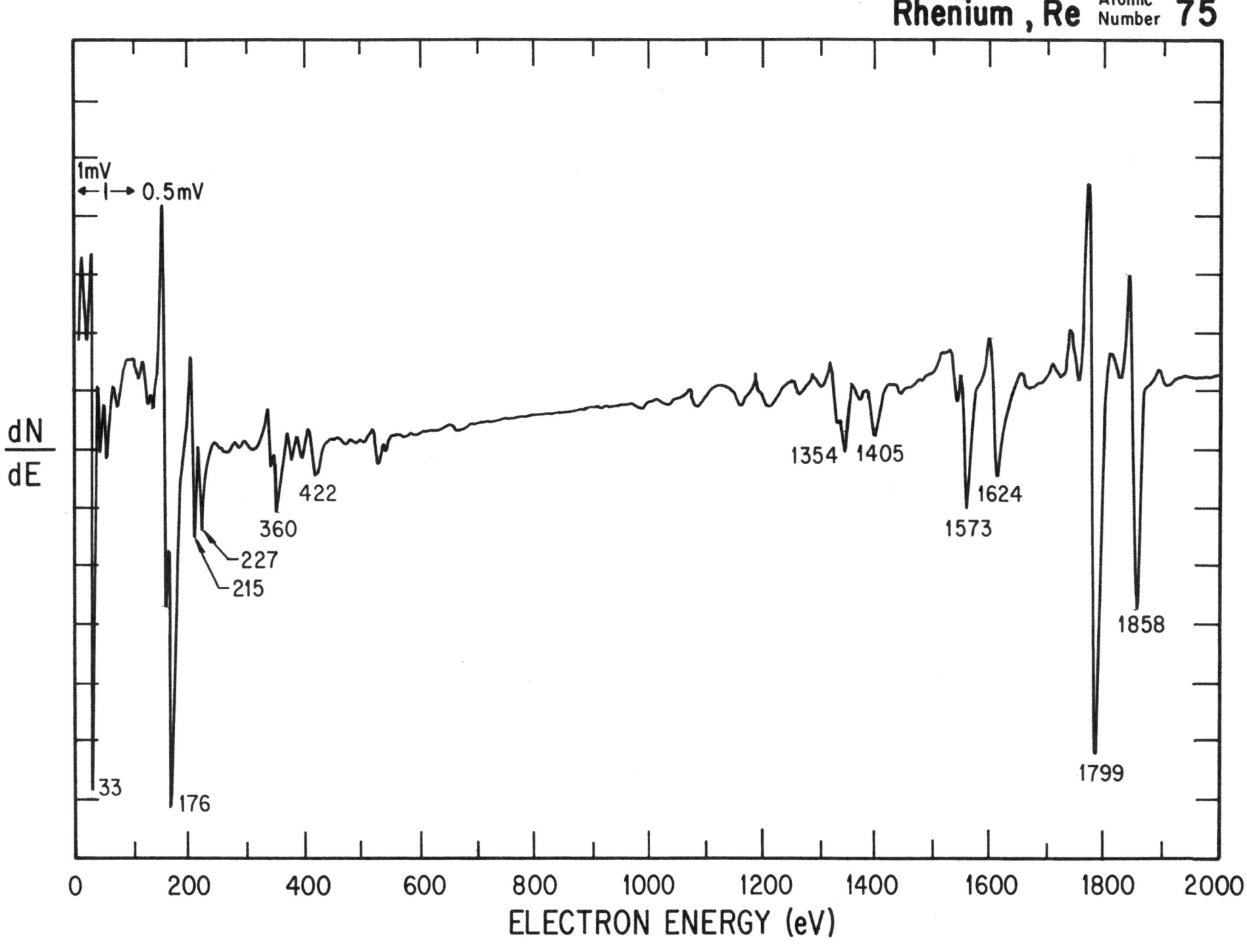

Rhenium, Re ᴬᵗᵒᵐⁱᶜ Number **75**

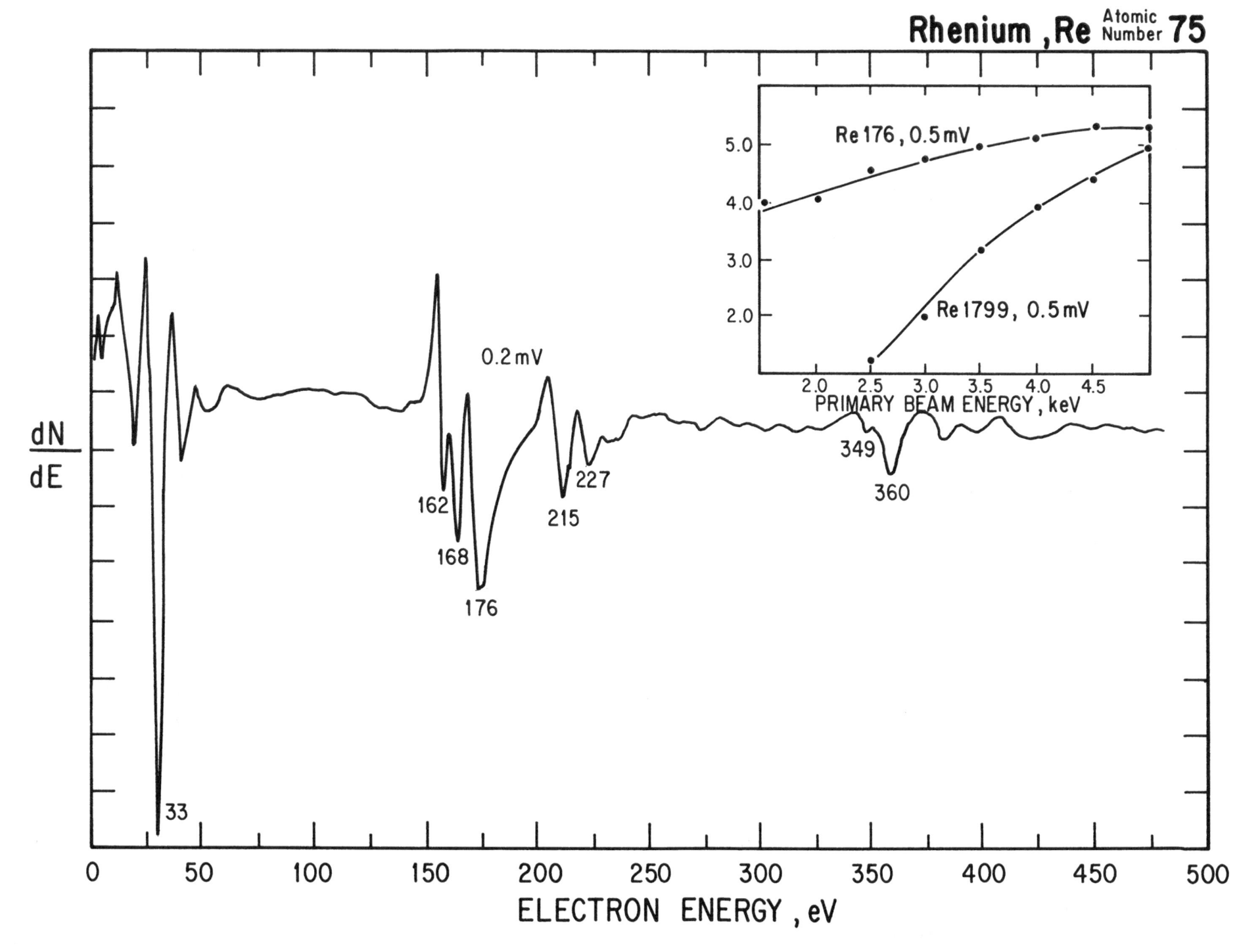

Rhenium, Re Atomic Number 75

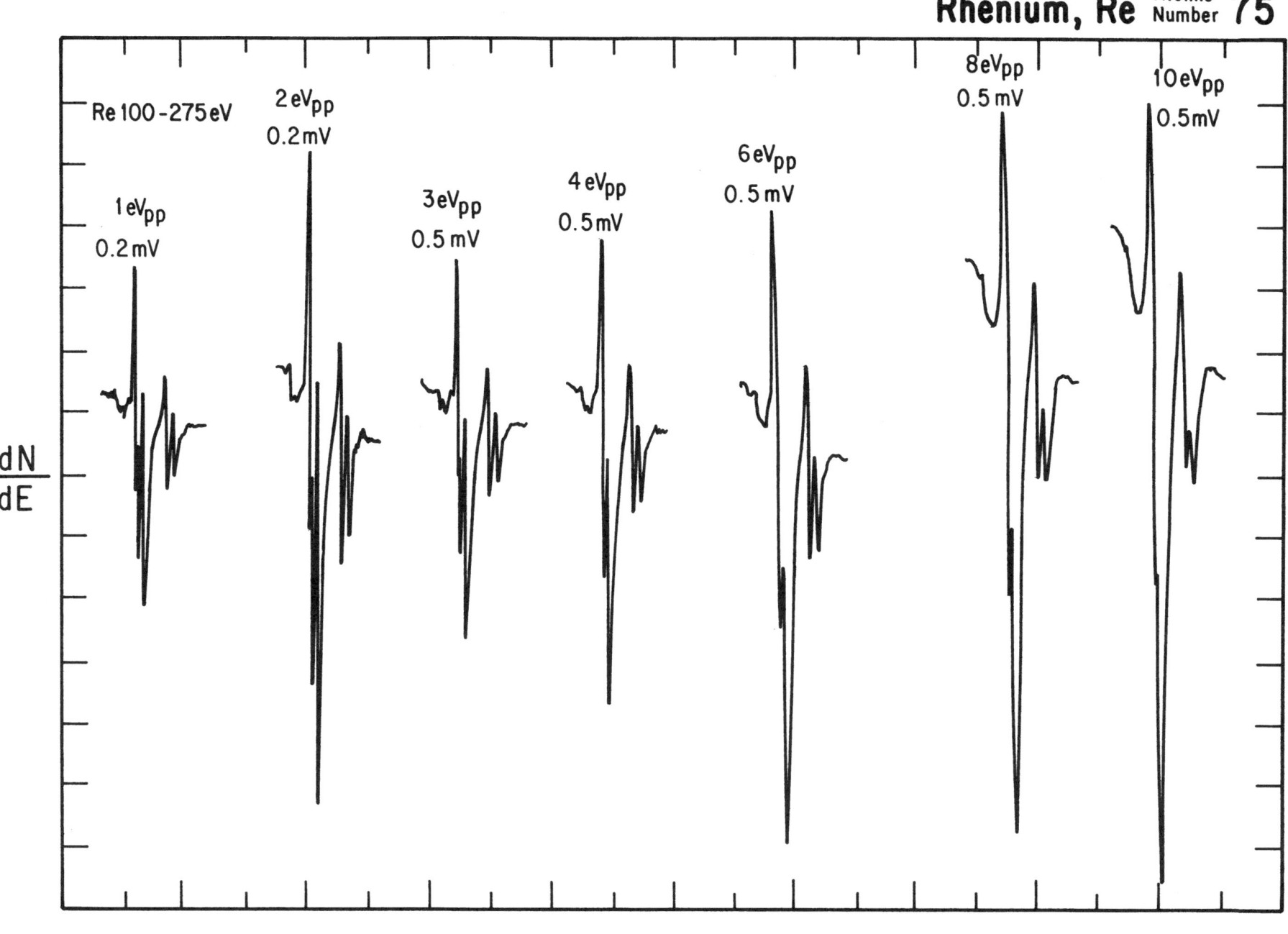

Re 100-275 eV

1 eV$_{pp}$
0.2 mV

2 eV$_{pp}$
0.2 mV

3 eV$_{pp}$
0.5 mV

4 eV$_{pp}$
0.5 mV

6 eV$_{pp}$
0.5 mV

8 eV$_{pp}$
0.5 mV

10 eV$_{pp}$
0.5 mV

$\dfrac{dN}{dE}$

ELECTRON ENERGY, 200 eV/DIVISION

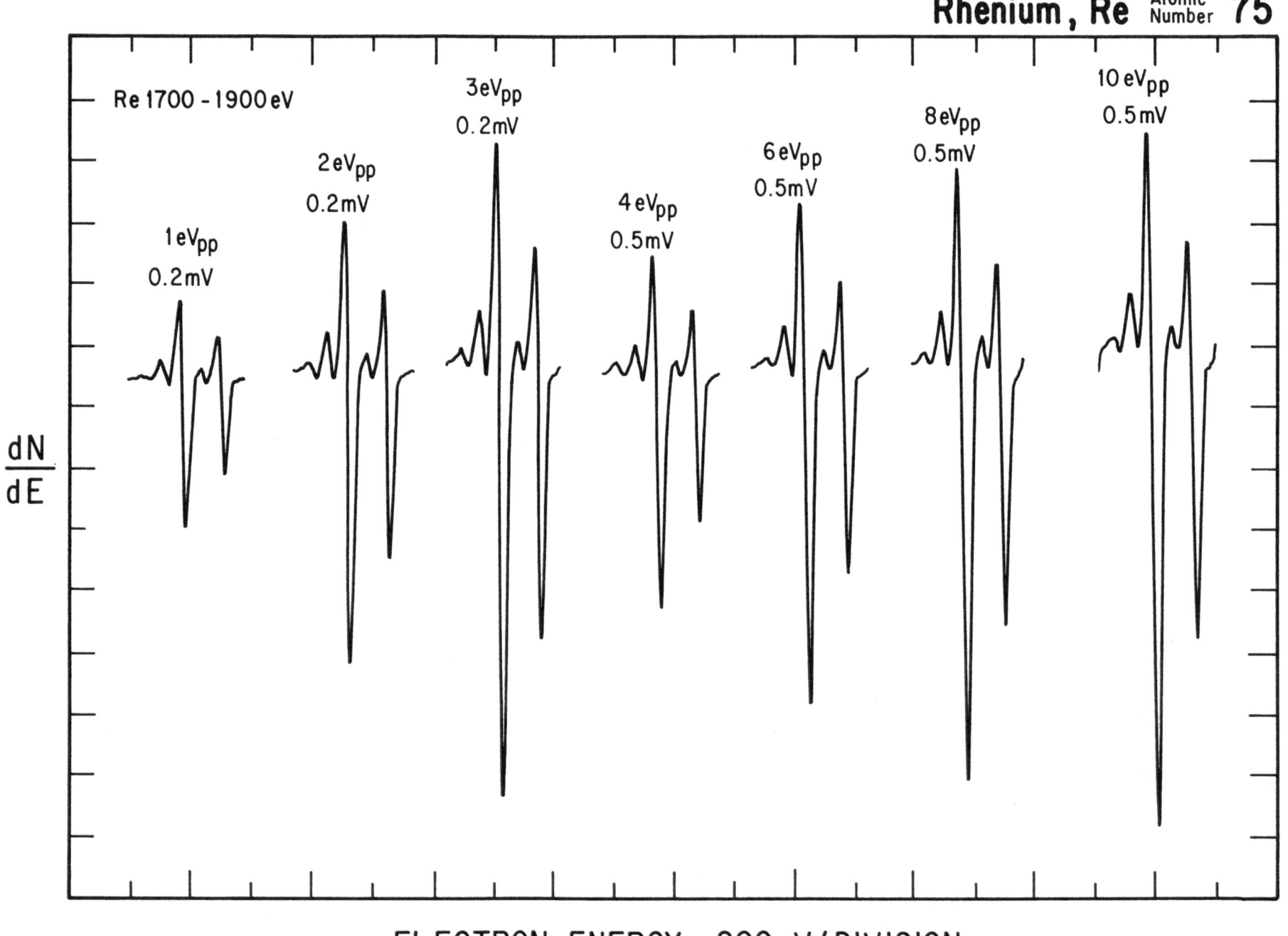

Rhenium, Re Atomic Number 75

Re 1700 – 1900 eV

1 eV$_{pp}$
0.2 mV

2 eV$_{pp}$
0.2 mV

3 eV$_{pp}$
0.2 mV

4 eV$_{pp}$
0.5 mV

6 eV$_{pp}$
0.5 mV

8 eV$_{pp}$
0.5 mV

10 eV$_{pp}$
0.5 mV

$\dfrac{dN}{dE}$

ELECTRON ENERGY , 200 eV/DIVISION

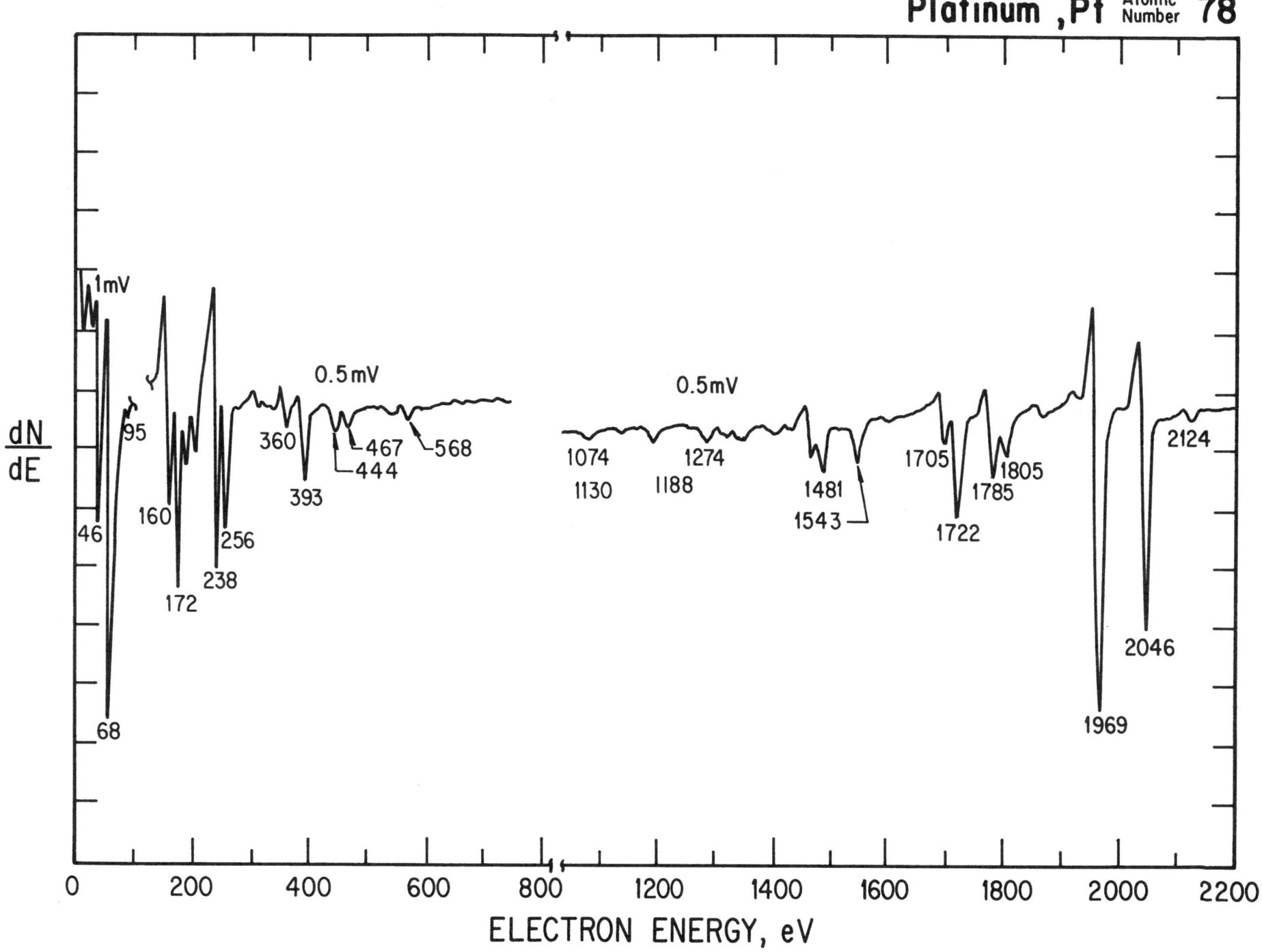

Platinum, Pt Atomic Number 78

dN/dE

ELECTRON ENERGY, eV

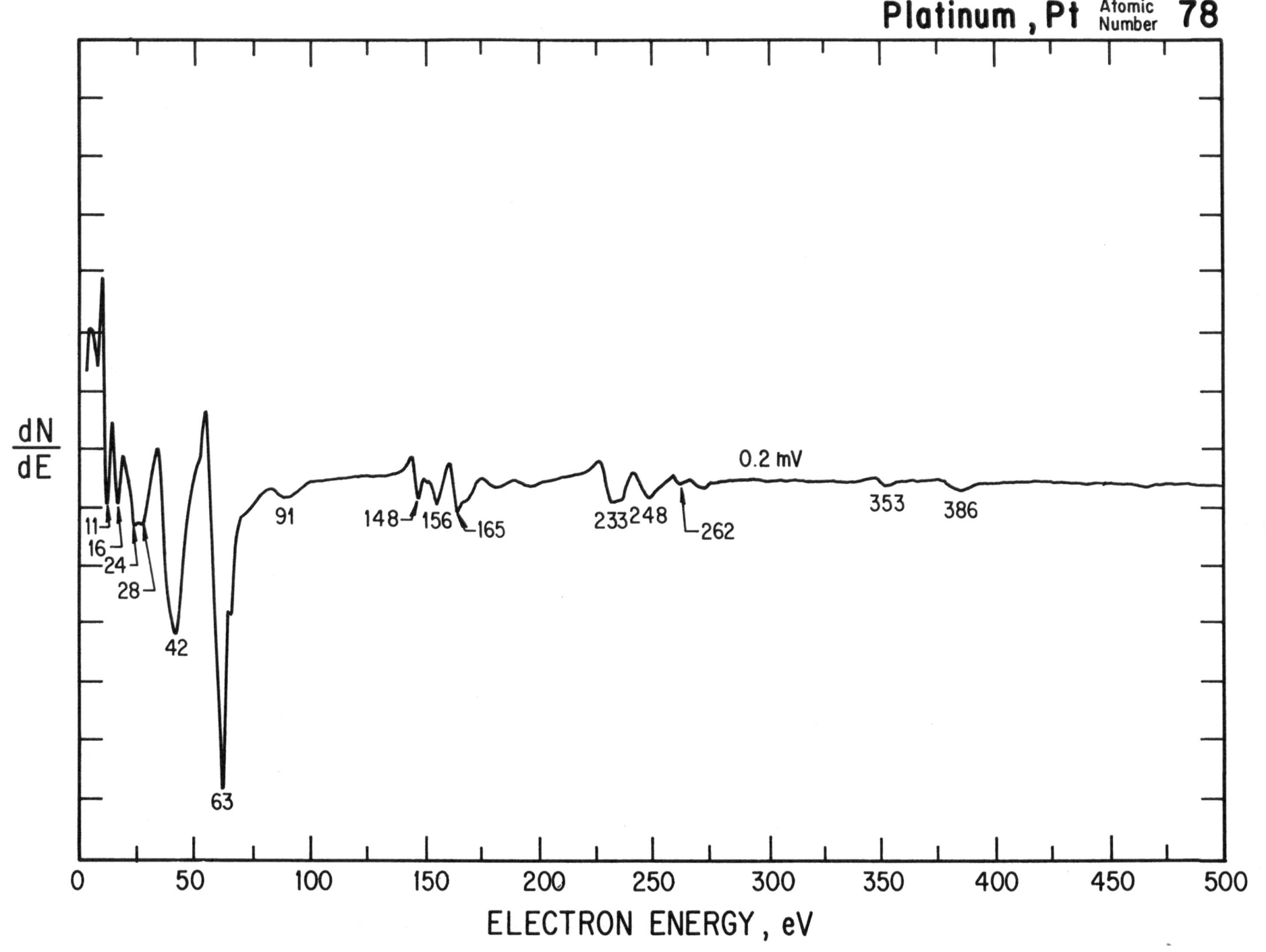

Platinum, Pt Atomic Number 78

dN/dE

ELECTRON ENERGY, eV

11
16
24
28
42
63
91
148
156
165
233
248
262
353
386
0.2 mV

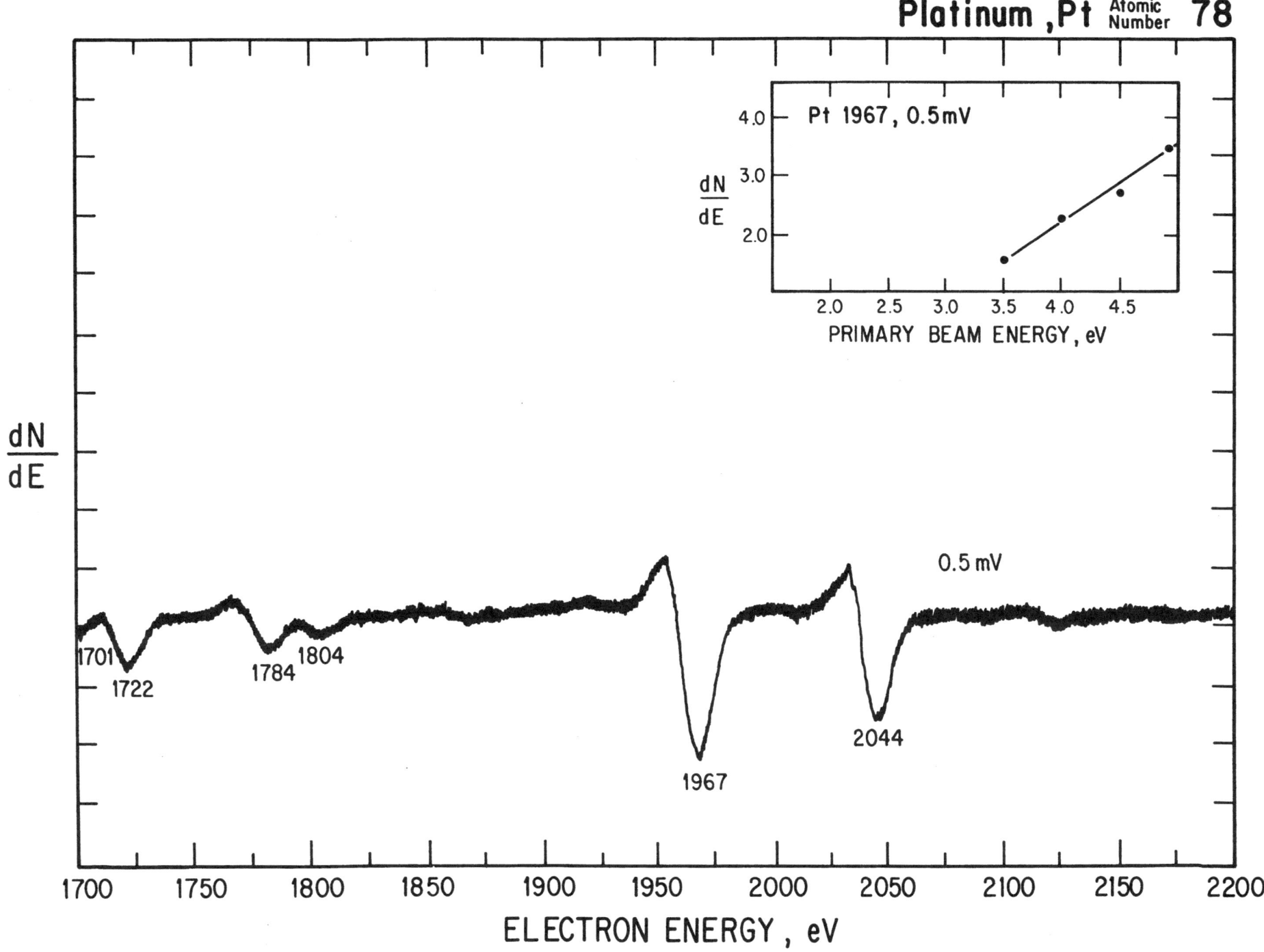

Pt 1967, 0.5mV

$\frac{dN}{dE}$

PRIMARY BEAM ENERGY, eV

$\frac{dN}{dE}$

0.5 mV

1701
1722
1784 1804
1967
2044

ELECTRON ENERGY, eV

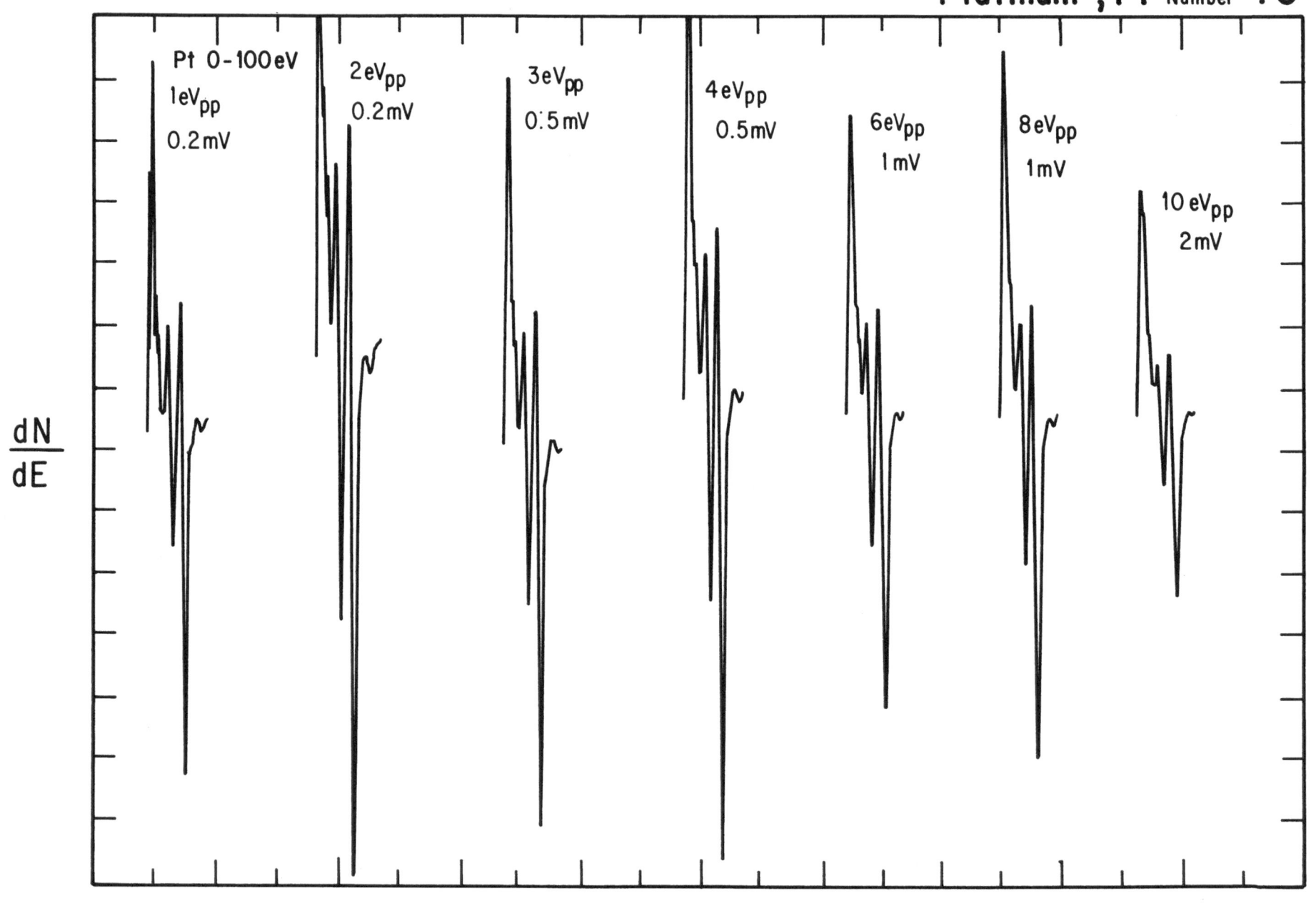

Platinum , Pt Atomic Number **78**

Pt 0-100 eV
1eV_pp
0.2 mV

2eV_pp
0.2 mV

3eV_pp
0.5 mV

4eV_pp
0.5 mV

6eV_pp
1 mV

8eV_pp
1 mV

10 eV_pp
2 mV

$\dfrac{dN}{dE}$

ELECTRON ENERGY , 200 eV / DIVISION

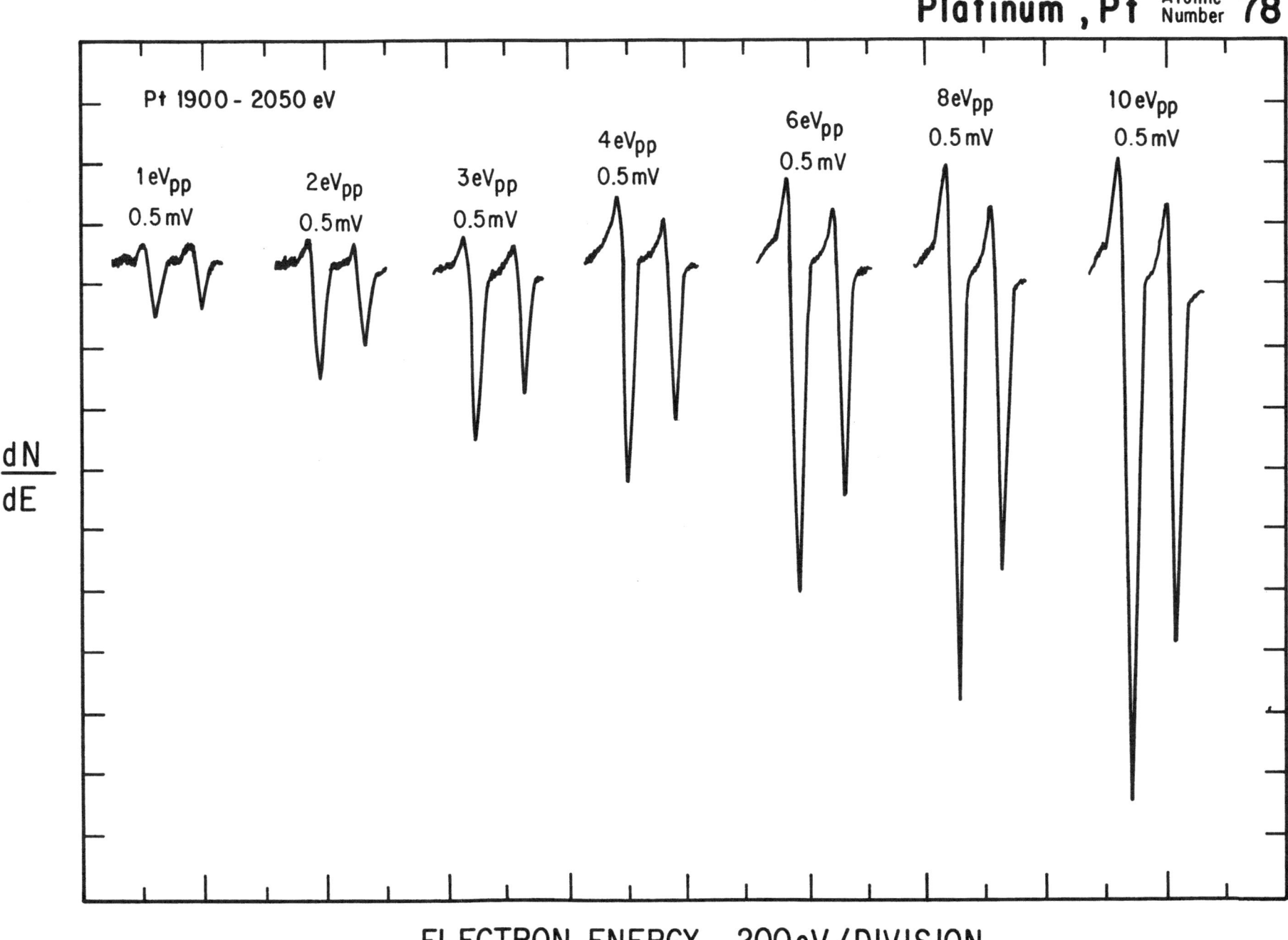

Pt 1900 - 2050 eV

1 eV$_{pp}$
0.5 mV

2 eV$_{pp}$
0.5 mV

3 eV$_{pp}$
0.5 mV

4 eV$_{pp}$
0.5 mV

6 eV$_{pp}$
0.5 mV

8 eV$_{pp}$
0.5 mV

10 eV$_{pp}$
0.5 mV

$\dfrac{dN}{dE}$

ELECTRON ENERGY , 200 eV / DIVISION

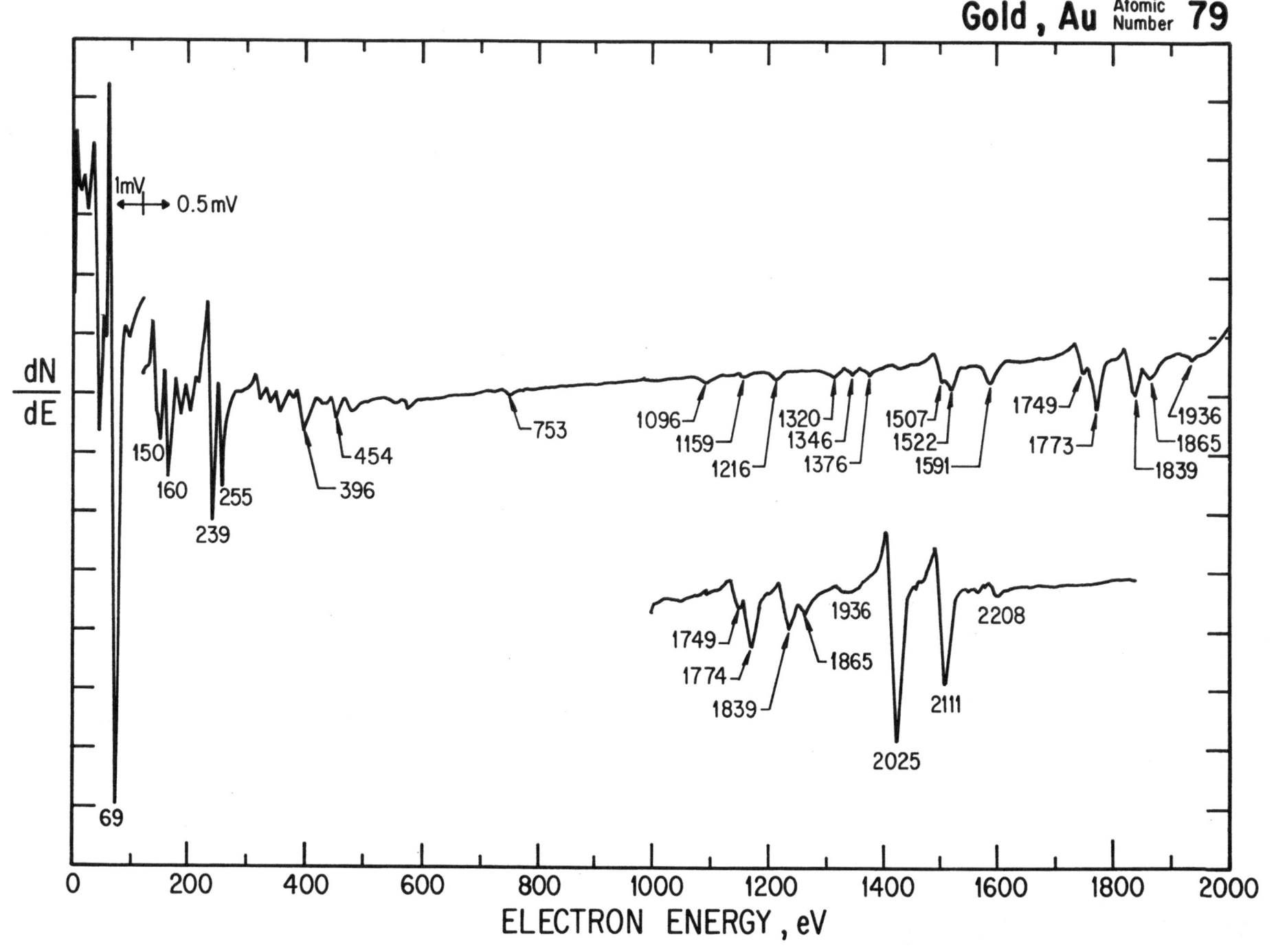

Gold, Au Atomic Number **79**

ELECTRON ENERGY, eV

$\dfrac{dN}{dE}$

1mV ← → 0.5mV

69

150
160
239
255

396
454

753

1096
1159
1216
1320
1346
1376

1507
1522
1591

1749
1773
1839

1865
1936

1749
1774
1839

1936
1865

2025
2111
2208

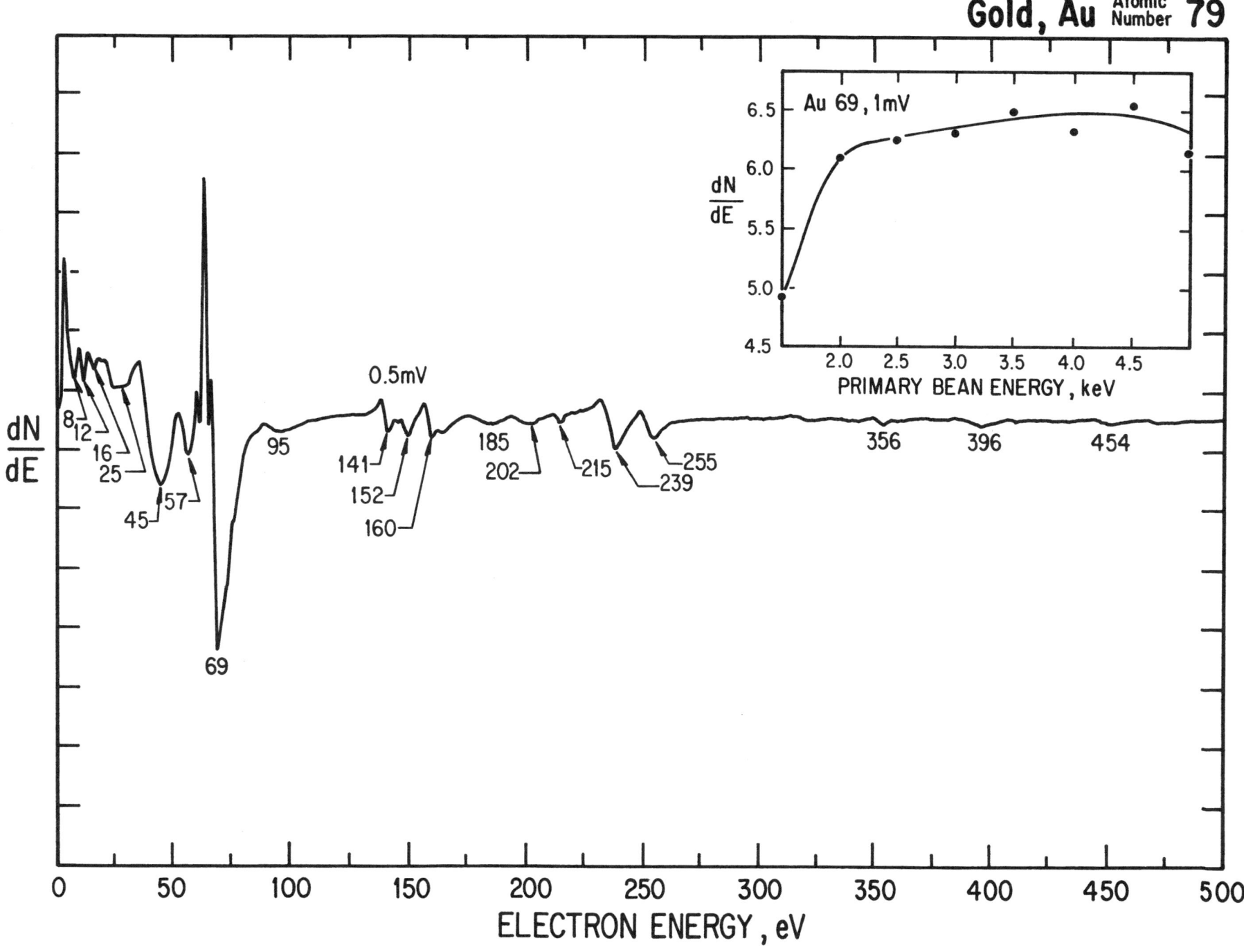

Gold, Au Atomic Number 79

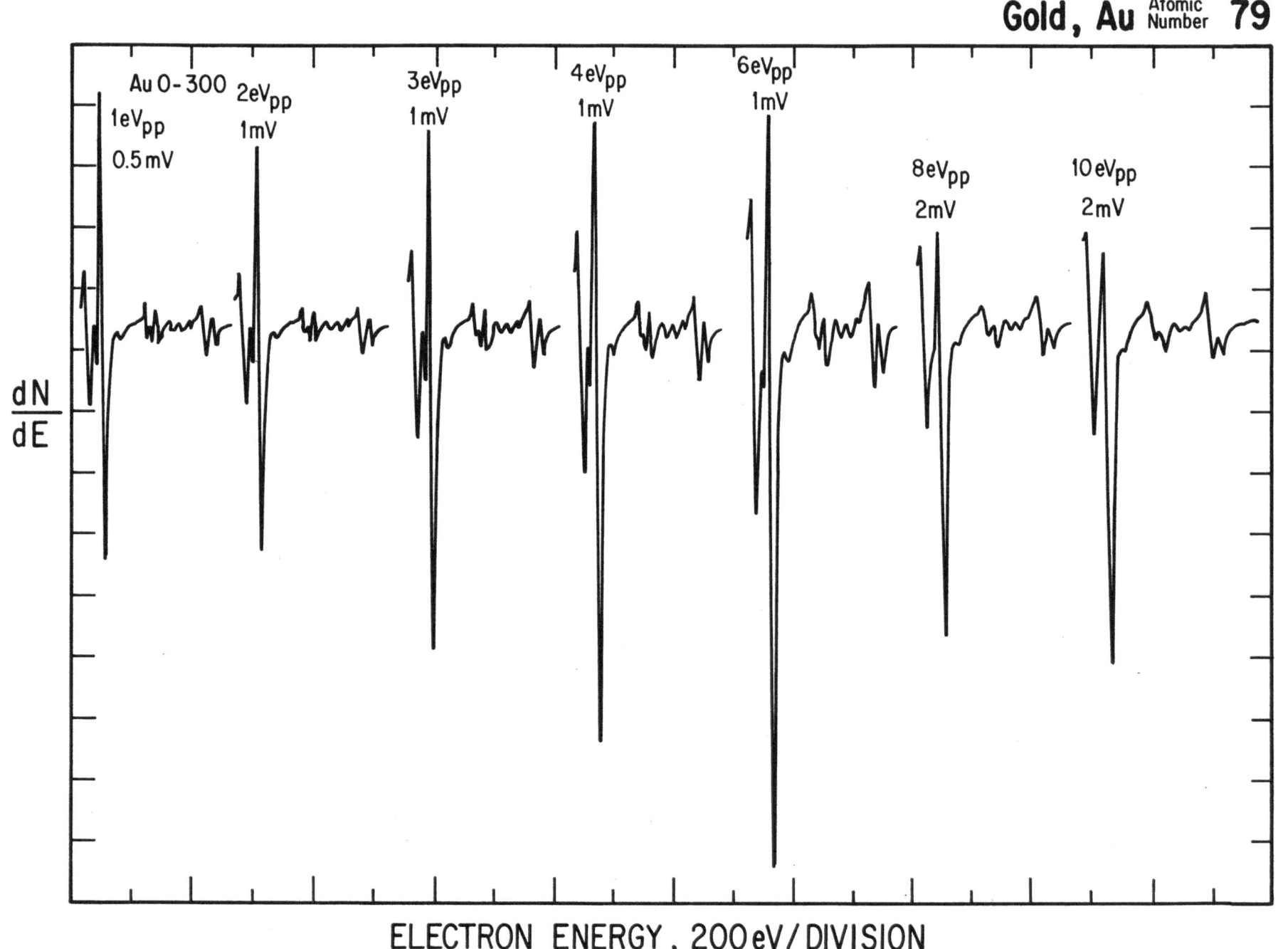

Au O-300

1eV_pp
0.5 mV

2eV_pp
1mV

3eV_pp
1mV

4eV_pp
1mV

6eV_pp
1mV

8eV_pp
2mV

10eV_pp
2mV

$\dfrac{dN}{dE}$

ELECTRON ENERGY, 200 eV / DIVISION

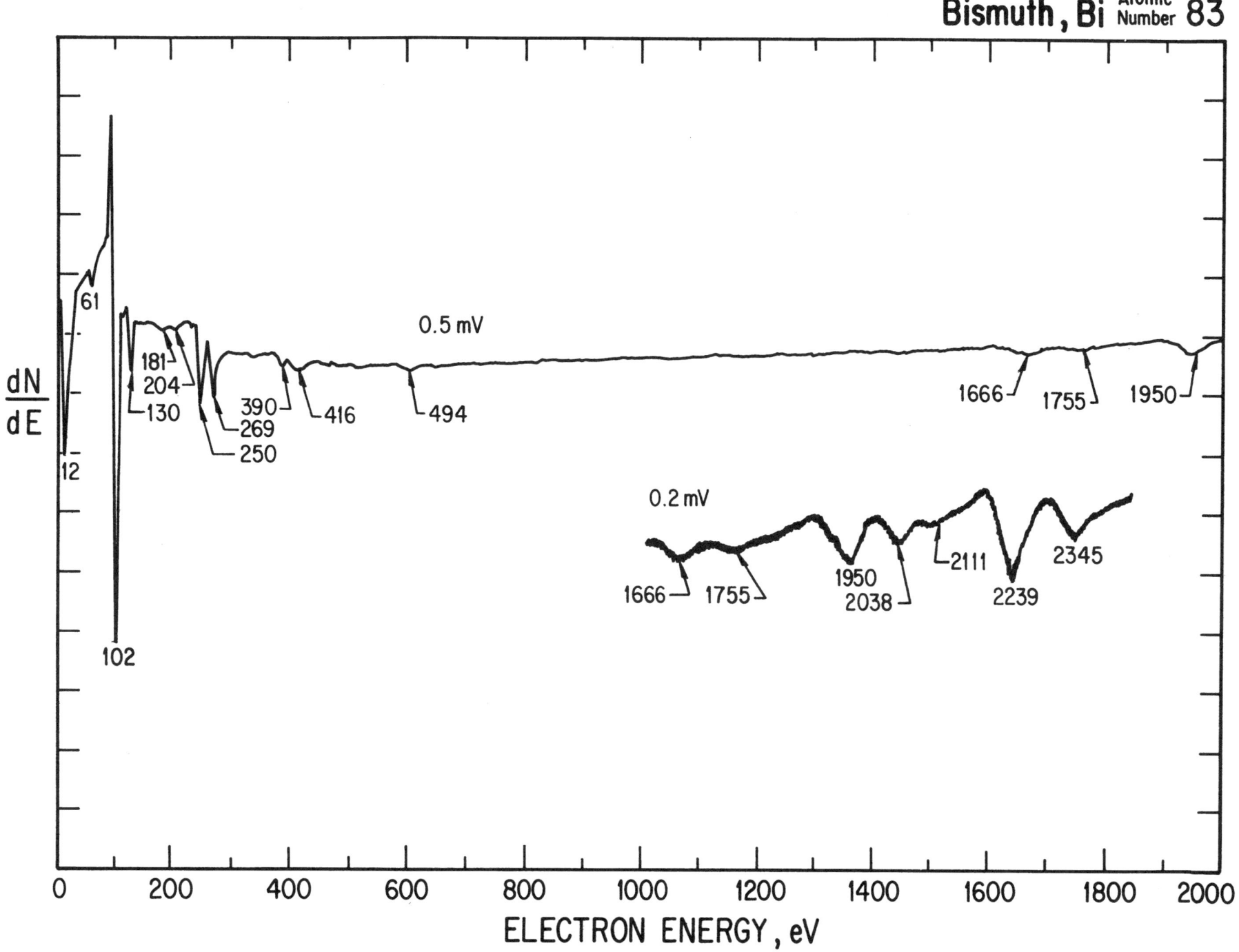

$\dfrac{dN}{dE}$

ELECTRON ENERGY, eV

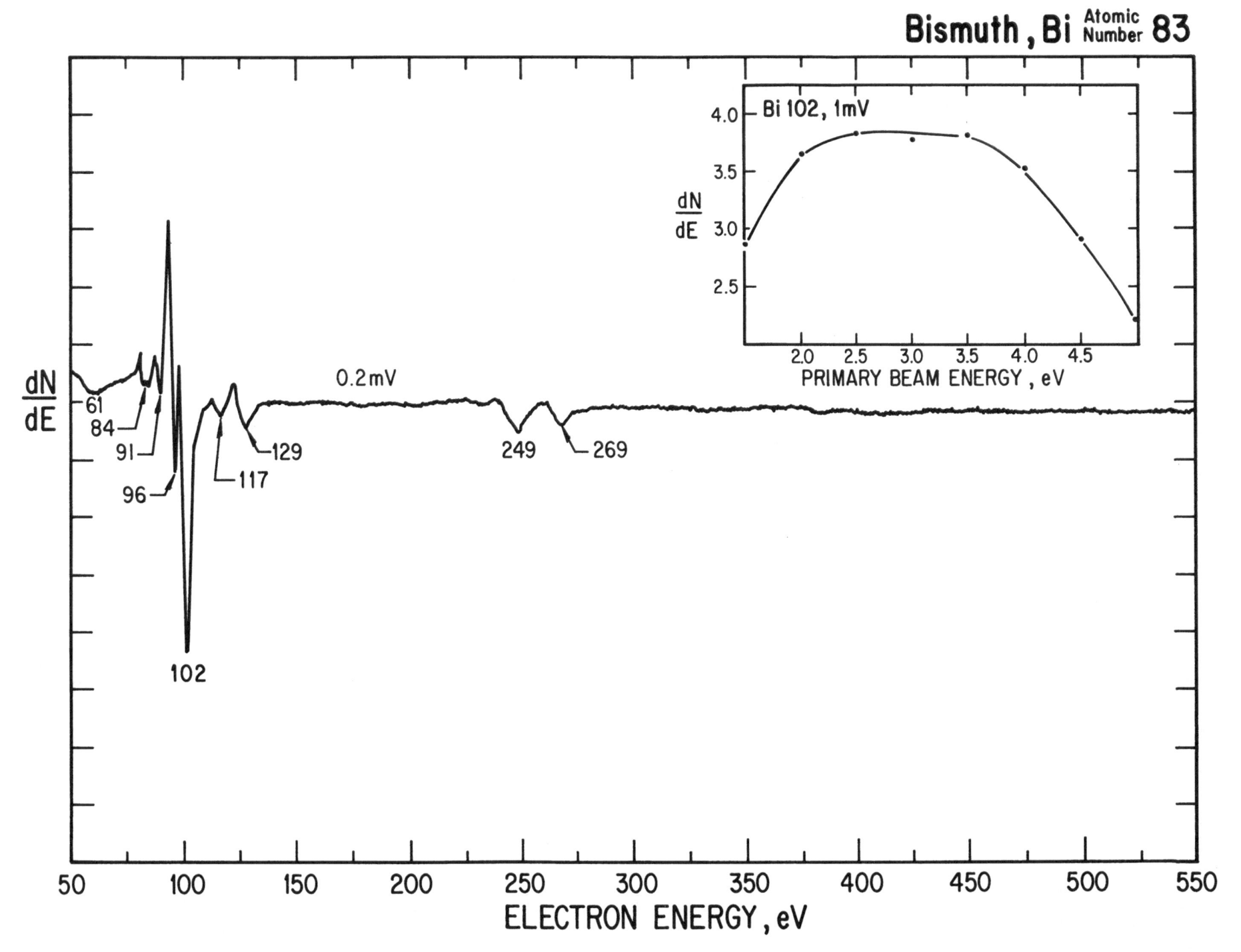

Bismuth, Bi Atomic Number 83

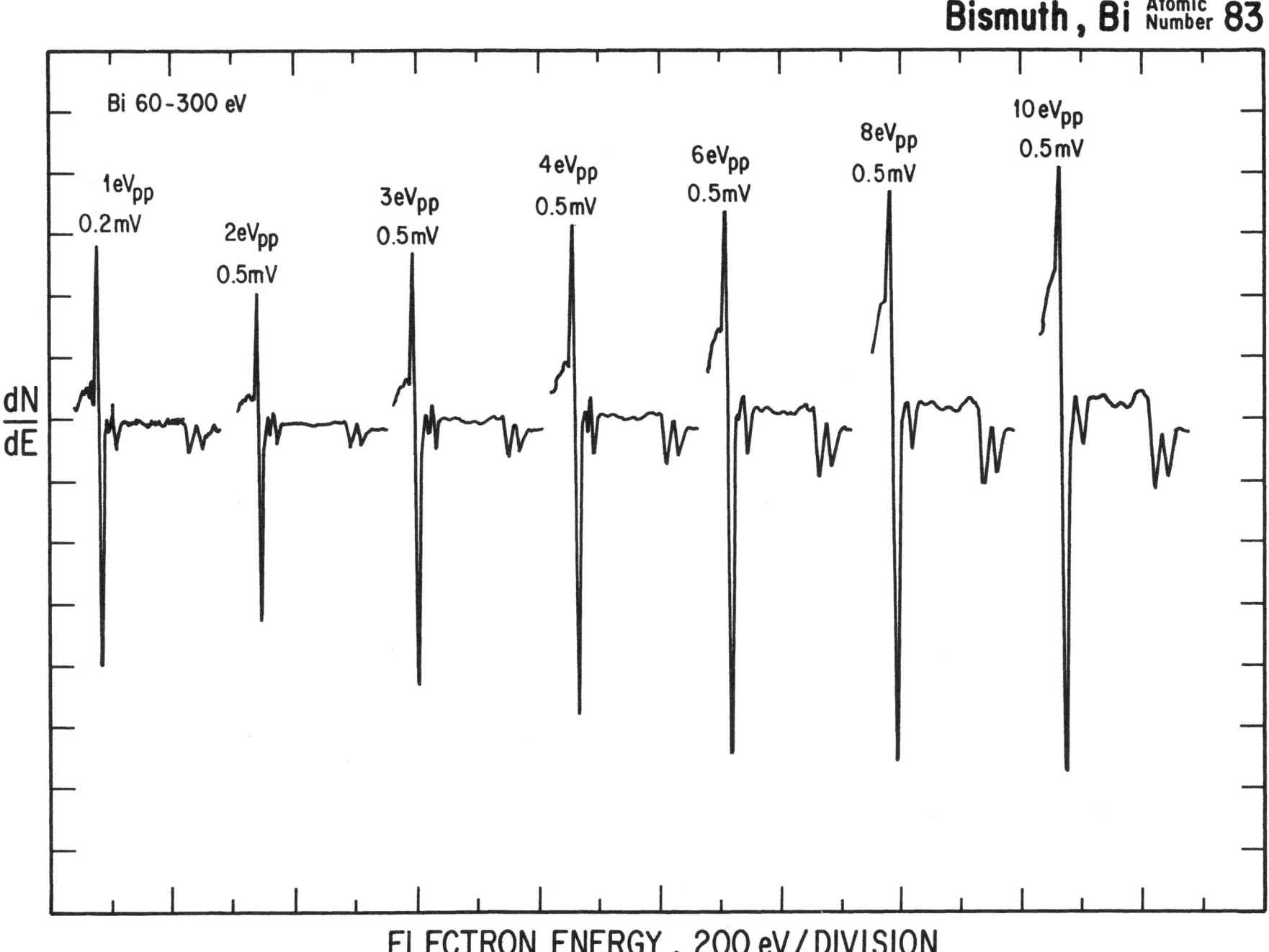

Bi 60-300 eV

1eV_pp
0.2mV

2eV_pp
0.5mV

3eV_pp
0.5mV

4eV_pp
0.5mV

6eV_pp
0.5mV

8eV_pp
0.5mV

10eV_pp
0.5mV

$\dfrac{dN}{dE}$

ELECTRON ENERGY , 200 eV / DIVISION